ENVIRONMENTAL MANAGEMENT

ENVIRONMENTAL MANAGEMENT

Editor
Prakash Talwar

ISHA

ISHA BOOKS
DELHI-110033

Environmental Management

Rs. 650

ISBN: 81-8205-341-2

Published in 2006 in India by
Isha Books
D-43, Prithviraj Road,
Adarsh Nagar, Delhi-110033
e-mail : ishabooks@hotmail.com

Laser Typesetting at : Shiva Infosystem, Delhi
Printed at : Mehra Offset Press, Delhi

Preface

The objective of environmental management is improved human life quality. It involves the mobilisation of resources and the use of government to administer the use of both natural and economic goods and services. It is based on the principles of ecology. It uses systems analysis and conflict resolution to distribute the costs and benefits of development activities throughout the affected populations and seeks to protect the activities of development from natural hazards. In the complex and interdependent world that we have been given, environmental management is required because the activities of development in one sector affect in both positive and negative ways the quality of life in others.

Designed as a guide to hands-on, day-to-day environmental management, this book covers the nuts and bolts of effective environmental management. It provides a thorough introduction to those new to the field or to environmental specialists who need a thought-provoking refresher. It will enable the students to gain a practical understanding of environmental hazards, controlling air and water pollution, managing hazardous wastes, implementing pollution prevention programs, dealing with compliance audits and regulatory inspections, and much more.

Editor

Contents

1

Introduction to Ecology

Ecology is the study of animals and plants in their relations to each other and to their environment. German biologist, Ernst Haeckel was coined the term oekologie (ecology) in 1866. The Greek oikos meaning "house" or "dwelling", and logos meaning "science" or "study". Thus, ecology is the "study of the household of nature". Haeckel intended it to encompass the study of an animal in relation to both the physical environment and other plants and animals with which it interacted. Ecology is regarded as multidisciplinary so broad is its potential scope.

Scope of Ecology

Mankind has been studying nature for thousands of years and formally for several centuries under the science of Biology. Ecology is one of three main divisions of biology; the other two being morphology [organism structural aspects] and physiology [organism functional aspects]". The behavioral sciences encompass one facet of the interaction of an organism with its physical and biological surroundings, and therefore are part of ecology.

It is a fact that ecology incorporates and overlaps with many other disciplines in both the biological and physical sciences. Certainly on one level, there is no information about the natural environment that does not have some applicability to ecology. On the other hand, such interrelationships between the sciences are much more the norm than most people realize. Ecology, is a distinct science because of the way information about the natural world is organized; and because of its unique methodologies and point-of-view by which knowledge is extracted from such information.

Ecology has no aim, but ecologists have. The problems of the ecologist are not fundamentally different from those of any other kind of naturalist. The superficial differences in aim are due to the different points of view, or methods of approach, rather than to any essential differences in the character of the problems.

Ecologists study in:

— structural, functional, and behavioral adaptations of organisms in relation to their environment;

— interrelationships between species and their populations, and populations in their communities;

— local and geographic distributions of organisms;

— regional variations in organism abundances;

— temporal changes in the distribution, abundance, and behaviors of organisms; and

— evolution of the interrelationships described.

Ecology is both a biological and an environmental science. Many environmental sciences are minimally concerned with biology (meteorology, for example) and others (environmental toxicology, for example) necessarily combine physical and biological sciences.

Biosphere and its Units

Ecology is essentially the study of the workings of the

planet earth. As ecology took hold among biologists in the 19th and early 20th centuries, many concepts of ecological organization were proposed. These came from both zoologists and botanists-whom seldom read the other discipline's literature-and from different schools of thought in (mostly) America and Europe.

However, ecology is not a course that can be persued without first acquiring a reasonably strong background in biology and it is assumed that the student has accomplished this. You will need to understand the conceptual basis of biological species, organism genetics, and evolution. You should quickly recognize, as the lessons proceed, whether your training in biology is in fact sufficient.

The term environment describes, in an unspecified way, the sum total of physical and biotic conditions that influence an organism. The subset of the planet earth environment into which life penetrates is termed the biosphere. With respect to the planet earth, the biosphere penetrates only a limited distance into the rock beneath the land and the oceans, and a limited distance out away from the planet towards space. All human effort so far has failed to demonstrate that the biosphere extends beyond these limits or that other biospheres exist elsewhere in the universe.

We cannot therefore conclude that they do not exist, only that we know nothing of that existence. The distinction is an important one to grasp, because we often encounter unexplained phenomenon ascribed to "extraterrestrial life" .

Ecosystem is perhaps the most widely used term in ecology and a concept basic to what ecologists do. The study of animals and plants in their relations to each other and to their environment; ecosystem is the system of organisms and physical factors under study or consideration. Although the boundaries of ecosystems are sometimes quite difficult to define in nature, ecosystems-however bounded-comprise the basic units of that nature.

The basic physical units of the biosphere are the lithosphere (the land), hydrosphere (the water), and atmosphere (the air). Apparently there is no permanent biota of the atmosphere, although insects and birds among others utilize that environment extensively. These basic units are easily recognized in any landscape, as for example shown in the scene at right from Belarus in Europe.

Ecologists regard ecosystems as basic structural units, implying that the biosphere is a collection or mosaic of definable ecosystems. However, an ecosystem is really more conceptual than actual. By this we mean that the definable boundaries of an ecosystem are more in the mind of the ecologist than geographical. We can certainly observe that there are many boundaries in nature-some fairly sharp-and what we are observing are physical manifestations between volumes wherein different conditions prevail.

These are biocycles, subdivisions of the biosphere. Consider a lake shore, or even the water surface for that matter; the transitions between the biocycles-land, the air or atmosphere, and the body of water-are sharp and recognizable upon casual inspection. There is no doubt that under gravity, water in the liquid phase sorts itself out from gaseous air-a lake surface exists. But along the shore, the water moves into the earth wherever the soil or voids in the rock allow.

Constantly water-saturated soil becomes a wetland. Further, atmospheric gases dissolve in the water of the lake, and molecules of water move off the lake as gaseous moisture, evident perhaps as fog on a cool morning. The sharpness of the boundaries is both actual and illusionary. Actual, because conditions encountered within each of the separate spaces or volumes (air, land, water) are very different. Illusionary, because no biocycle is completely isolated from adjacent biocycles, and typically each has influence over the other.

We have, so far considered just the physical environment. Ecosystems include organisms. And we can anticipate that most, though certainly not all, organisms will sort themselves out along the same physical boundaries described. That is, fishes will find life conditions suitable only in the lake; conifers only in the uplands. Yet organisms are no more isolated from adjacent biocycles than are the substances of the physical world. The food of many fishes will be insects living only on the land and moving through the atmosphere, which upon death or by accident, fall into the water.

The conifers draw up moisture from the earth, some directly from the lake or the groundwater body connected to the lake. Ecologists use the term biotope to mean a topographic unit characterized by a particular set of physical conditions and a uniform assemblage of plant and animal species. Considering example above, the lake is part of the major freshwater biocycle and would be a biotope of the region in which it is located, being perhaps just one of many ponds and lakes comprising that biotope. A biochore is a subdivision of the biosphere consisting of the biotopes that resemble one another; thus, lake is but one example of the temperate zone, freshwater, lake biochore.

The terms "biotope", "biochore", and "biocycle" fill a need in descriptive ecology. We can speak in a general way about the life forms and climatic conditions that constitute a desert biochore, or a northern forest biochore.

Gaia Theory

Consider the importance of the functional relationships

between species comprising an ecosystem, and how those relationships contribute to the very survival of the ecosystem. Consider that ecosystems are not, in most cases, isolated units, but parts of a larger functioning whole, at its broadest being the biosphere.

Species and Habitat

In biology, the concept of species is basic. A species comprises a group of related organisms that share a more or less distinctive form and are capable of interbreeding. Species are groups of actually or potentially interbreeding natural populations which are reproductively isolated from other such groups.

In biology, it is generally the case that focus on the species concept is largely the concern of taxonomy and taxonomists. In ecology, the more important aspect is that of the species population. In ecology we are less concerned with all considerations of what "potentially interbreeding" might mean and more interested in the group of individuals that are, in fact, interbreeding.

By which is meant, not that sexual reproduction must take place between all extant (living at the same time) members, but that there exists a gene pool that is open and potentially contributed to by each member. The concept of species would admit all those individuals separated by great distances and physical barriers as included if breeding could still take place. The species population takes into account the isolation that might actually exist, frequently dividing a single species into more than one population.

Within the population, alleles presumably flow freely through time, held (at any given point in time) within the nuclei of the cells of the individual organisms constituting the population. The term, habitat, has a number of different uses in English and in biology, but is generally considered by biologists to be the physical conditions that surround a species, or species population, or assemblage of species, or community.

Many ecologists extend the term habitat in the manner originally defined to cover physical parameters relative to a community of species, but others would substitute the term "biotope" for "habitat" in such cases. Note that this is not exactly the same as "biotope" defined above, which included the biota as well as the physical or abiotic environment. Any term in ecology that attempts to separate the biotic and abiotic components of an ecological unit of organisation will encounter this difficulty.

Consider that if the habitat of a bird is the place it lives, and the bird lives in a pine forest (actually up among the branches of the pine trees), then the physical environment (the "habitat") of that bird species must include a life form (pine trees); and now extend this reasoning to any manner of invertebrates, such as flea or tapeworm, that really do live most of their life on or in other organisms. A reasonably solid definition of habitat would be simply the environment of a species.

A niche or ecological niche is essentially what a species does. We do not mean so much what activities it participates in (although these could be pertinent), but the role the species plays in the functioning of the ecosystem: the "functional status of an organism in its community".

There is but one known biosphere, that of the planet earth. The biosphere can be subdivided into a number of biocycles, each with fairly specific gross physical characteristics and limited by geography and topography as determined by planetary dynamics, especially gravity. Biocycles are divided into biochores, units with both similar basic physical attributes and life forms. Biochores are made up of a multitude of biotopes, divisions of the landscape that feature both similar physical attributes and a more or less consistent species composition.

Some ecologists would use "biotope" for just the physical properties of this level of organisation, and biocoenosis for the biotic component. However, such definitional division of highly integrated and complex ecological systems is difficult to uphold. At the level of the species population, the habitat describes where it is found, where that species occurs in the landscape. The very different term, niche, describes what that organism does as a functional role in the ecosystem. The ecosystem concept differs from traditional (or the "old") biology by its clear emphasis on functional relationships among organisms.

Environmental Response

Basic to the interest of ecologists is how organisms interact with their environment. You can understand that inanimate objects do not normally "interact"-they are simply acted upon by forces in the environment. Organisms, however, cannot be totally passive within a chain of events and still sustain life. In a sense, life represents a reversal of the "normal" order of things in a universe where physical and chemical reactions proceed, on balance-and usually rather directly-towards lower energy states.

When a volcano erupts it lays down rocks on the surface of its slopes. These rocks are elevated in energy state from their previous existence as physical material deep within the earth. But on balance, in order to erupt, that material was first heated until fluid, and then tremendous amounts of energy expended (as pressure) to push the material (magma) up to spew as lava from the top of the mountain. Only a small portion of that energy remains on the volcano in the form of potential energy of rocks perched on the surface.

That potential energy will eventually be released after forces of erosion loosen the rocks and they roll downslope or are ground to silt and deposited in the ocean deep. In essence, the physical environment (as distinct from the biological environment) is composed of various forms of matter and energy. The physical environment is one of material acted upon by various forces (energy), resulting in a lower energy state of the material with a concomittant release or conversion of energy as (usually) heat or light.

Living systems-organisms-respond in various ways to contacts with the physical forces of an ever-changing environment and interrelationships with other living organisms. The hereditary potentialities of an organism determine what it can do, but the environment determines what it actually does and to what degree. Organism responses can be described under four different categories:

— morphological,

— physiological,

— behavioural, and

— community relations.

An organism's response to an environmental challenge seldom involves just one of these. At the very least, the growth or development of a morphological feature involves internal physiological processes occurring between the genetic blue-print in the cells and some final experession of an anatomical trait. A protective morphological trait like horns on a bull come into play through behavioral actions. The horns are not a threat to a predator until wielded in an effective manner.

Organism Morphology

The form that an organism takes-termed its life-form-encompasses all of the structural aspects that make that species describable and unique. By morphological adaptations is meant the structural feature or features (called traits) that allow an organism to live successfully in its habitat; traits that have evolved over a (usually long) period of time as a consequence of the process of natural selection.

Adaptive morphological traits are acquired by a species slowly through the evolutionary process and therefore not subject to change within the lifespan of an individual, in fact many morphological traits are expressed or modified in response to environmental conditions. The individual organism inherits morphological variation within a range, and the expression of the trait represents a response to specifics of that individual's environment.

The considerable similarity of all individuals of a species attests to the genetic basis of morphological traits, yet a degree of plasticity is usually present to be influenced by environmental factors. Individuals inherit this plasticity and not the growth-form of the parents. Thus growth-form represents the outcome of both "nature" (genetic or inherited traits) and "nurture" (influences of environment on the growth process). A growth-form that persists through many generations because of environmental constancy (but is not an inherited trait) is termed an ecad.

However, any such trait is subject to natural selection and may, over time, become fixed in a population as an inherited eco-type. That is, if the trait is highly adaptive within the habitat of the population, and the plasticity of expression is under-utilized becoming a genetic burden, that plasticity may be lost or restricted. Eco-types are genetically based differences between individuals of the same species that come from different populations.

Organism Physiology

Physiology is the study of biochemical functions of living

organisms. Whereas morphology is essentially about anatomy and structure, physiology is largely about processes. However, the two are so interlinked in any organism that one cannot really be discussed without consideration of the other. As already noted, physiological processes must procede the growth of morphological traits. And a physiological response, such as an increase in blood preasure, results from short-term mechanical changes in the anatomy of the heart and arteries.

Put another way, the shape and structure of a heart is a morphological trait; the pumping of blood-the function of the heart organ-is a physiological activity. Muscle contractions are mechanical actions resulting from biochemical signals and changes. One cannot occur without the other.

Biologists refer to physiological adjustment as a response to environmental change. By "adjustment" is meant a change in physiology or internal function that allows the organism to maintain life. An adjustment is limited by the physiological and morphological adaptations inherited by the organism. Just as previously described "morphological variation" as essentially an adjustment to environment based upon inheritance of a range of possibility, so too is physiology controlled by both inherited (genetic) and environmental (external) factors.

The important distinction here is between the terms "adjustment" and "adaptation", a distinction that can apply to both morphology and physiology.

— Adaptations are physiological processes or morphological traits inherited through genetics and subject to natural selection at the population level;

— Adjustments are activities and traits expressed within an inherited range and subject to individual modification within the life-span of the individual or even much shorter periods of time. The proper term here is acclimatization.

On the other hand, when a caterpillar changes into a pupa and then transforms itself into a butterfly, that is clearly an adaptation. The process (called metamorphosis) is an inherited one, and irreversable. Aging is a process that would appear to be largely irreversible (certainly death is), and apparently is an adaptation. Once reproductive success peaks and starts to decline, advantage for the species population goes towards eliminating older individuals.

Organism Behavior

Behavioral responses to environmental factors can be observed as movements of one sort or another. We say that "behavior" is the way an organism "acts"-what activities it "does". Complex behaviors controlled by cognitive processes evolved from simple response behaviors, the understanding of which provides a basis for understanding how behavior confers survivability on an organism.

Such responses are generally rapid (and mediated through nerves in higher animals) in organisms capable of locomotor activity, and generally much slower in organisms lacking means of motility. Thus, plants, although devoid of musculature, nonetheless respond to stimuli of various sorts-growing or turning towards a light, for example. Simple turning responses, either towards or away from a specific stimulus, are termed tropisms. Actual physical movement towards or away from a stimulus are called taxes. Similar behaviours called kineses involve physical movements that appear random in orientation or direction, but the intensity of activity is determined by the intensity of the stimulus.

Biogeochemical Cycles

A biogeochemical cycle or inorganic-organic cycle is a circular path by which either a chemical element or a molecule moves through both biotic ("bio-") and abiotic ("geo-") compartments of an ecosystem. In effect, an element is chemically recycled, although in some cycles there may be places (called "sinks") where the element accumulates and is held for a long period of time. In considering a specific biogeochemical cycle, we focus on a particular element and how that element participates in chemical reactions, moving between various molecular configurations.

Of the 90-odd elements known to occur in nature, some 30 or 40 are thought to be required by living organisms. We will be considering only a few of these, mainly those utilized in fairly large quantities by living organisms. The principal elements of life are carbon, hydrogen, oxygen, and nitrogen. However, a number of others are certainly important to understand as well, notably phosphorus and sulfur.

Hydrologic Cycle

A very significant (on planet earth) molecule that cycles through ecosystems is the water molecule (H_2O), for the reason that life is so dependent on water as the medium of chemical reactions within cells. The water cycle in terms of the various states of water, at least some water molecules are taken up by plants and split apart during photosynthesis into atoms of hydrogen and oxygen; the latter is released into the atmosphere as molecular oxygen (O_2).

Thus, through photosynthesizing organisms, the water cycle is an important part of both the oxygen and the hydrogen cycles. Note that the hydrogen ends up as part of an organic molecule, and therefore a significant participant in the carbon cycle. The majority of water in the water cycle is found within the oceans and the polar ices, although water is present in the bodies of organisms, in freshwater lakes and rivers, frozen in glaciers, and in the ground as groundwater. Water moves more or less freely between the storage reservoirs: by evaporation, precipitation, and runoff.

Carbon Cycle

Because carbon is required for the building of all organic compounds, carbon dioxide must be removed from the atmosphere and turned into a usable form. The reservoirs for the carbon cycle are the atmosphere, where carbon dioxide exists as a free gas, in fossil fuels and in durable organic material, such as cellulose. During the process of carbon fixation, carbon dioxide is absorbed from the reservoirs by plants, photosynthetic bacteria, and algae.

The carbon dioxide is "fixed" and it can then be used to construct organic substances. Animals receive their share of carbon by eating plants or other animals. Finally, the carbon cycle comes to completion when the "fixed" carbon is released when plants and animals undergo respiration, or when they decompose. The burning of organic material such as wood or fossil fuels can also result in the release of carbon dioxide.

Nitrogen Cycle

Nitrogen is required for the manufacture of all amino acids and nucleic acids, however the average organism can not use atmospheric nitrogen for these tasks and as a result is dependent on the nitrogen cycle as a source for its usable nitrogen. The nitrogen cycle begins with nitrogen stored in the atmosphere as N2 or nitrogen stored in the soil as ammonium (NH4+), ammonia (NH3), nitrite (NO2-), or nitrate (NO3-). Nitrogen is assimilated into living organisms through three stages: nitrogen fixation, nitrification, and plant metabolism. Nitrogen Fixation is a process which occurs in prokaryotes in which N2 is converted to (NH4+).

Atmospheric nitrogen can also undergo nitrogen fixation by lighting and UV radiation and become NO3-. Following nitrogen fixation, nitrification occurs. During nitrification, ammonia is converted into nitrite, and nitrite is converted into nitrate. Nitrification occurs in various bacteria. In the final stage, plants absorb ammonia and nitrate and incorporate it into their metabolic pathways. Once the nitrogen has entered the plant metabolic pathway, it may be transferred to animals when the plant is eaten.

Nitrogen is released back into the cycle when denitrifying bacteria convert NO3+ into N2 in the process of denitrification, when detrivorous bacteria convert organic compounds back into ammonia in the process of ammonification, or when animals excrete ammonia, urea, or uric acid.

Perfect Cycles

Odum describes what are called more or less perfect cycles: biogeochemical cycles that involve equilibrium states. That is, there exists in nature a balance in the cycling of the element between various compartments, with the element or material moving into abiotic compartments about as fast as it moves into biotic compartments. Certain ecosystems may experience "shortages", but overall a balance exists on a global scale.

ECOSYSTEM

Living organisms and their nonliving (abiotic) environment are inseparably interrelated and interact upon each other.

Any area of nature that includes living organisms and nonliving substances interacting to produce an exchange of materials between the living and nonliving parts is an ecological system or ecosystem.

But that activity is but a moment in time and cannot, by itself, define an ecosystem. The mouse must leave the burrow for food, interacting with plants that themselves interact and exchange materials with the soil and the atmosphere and are dependent in important ways on radiation from the sun and the activity of earthworms and fungi in the soil. The latter, in turn, feed upon organic molecules left behind from the death of former living plants.

It is the inclusion of all the interrelationships within a functional system that make the concept of ecosystem so broad. And the seemingly endless networks of links and interactions that give geographical breadth to an ecosystem. Of course, an ecologist could study just the organisms, activities, and biotic/abiotic interactions within the local "soil ecosystem", treating mouse and eagle poop as we described the sun above: external, but providing important input of energy and material to the soil ecosystem.

In a sense, this is what we mean by calling "ecosystem" a concept rather than just a place. Ecosystems over a wide range of sizes, from an aquarium to an ocean, and lots of places inbetween. In attempting to understand any interactive, complex system, especially one as elaborate as even a simple ecosystem, it is helpful to break the system down into various component parts.

By this mental process we can arrive at some understanding of the specific system and can make comparisons with other systems to further understanding of ecosystems in general. An ecosystem consists of many component structures and these function together in various ways, suggesting an initial approach to describe any ecosystem is to consider separately its structural and functional aspects.

Structural Perspective

Consider first an ecosystem from a structural perspective: an ecosystem consists of

— organsims,

— certain physical materials or resources, and

— the physical conditions or structure of the system that exist by virtue of its location on the planet.

We know or can observe that the organisms living in almost any ecosystem are not all identical individuals, but can be categorized into species; and each species has a unique set of morphological, physiological, and behavioral attributes that will determine how each individual functions within the whole. The species present provide ways of considering their contribution to ecosystem structure.

For examples: we could describe the number of species (species richness), or the number (or biomass) of individuals of each species (species diversity), or their distribution across the physical space of the ecosystem. These are a few ways an ecologist would approach describing ecosystem structure based upon the plants and animals (and others) present in it.

In addition to the organisms themselves, there are components utilized by the organisms that comprise the resources of the ecosystem. These are both biotic and abiotic, inorganic and organic, and are exchanged between organisms and between organisms and the environment. Resources are the materials cycled by the inhabitants of an ecosystem as they carry on with life processes.

Finally, there are those factors that are the conditions under which the organisms live. The environment of a subarctic lake is vastly different from that of a forest in the wet tropics. But consider also that conditions within a pond are very different from those just a few yards up from the shore. While we might be describing mostly physical conditions, these need not be always abiotic. In many ecosystems there are organisms that provide or create the dominant physical structure of that ecosystem.

Examples are a forest and a coral reef. In a forest ecosystem, the trees are both very prominant organisms and resources to many of the other species present; but they also have a substantial effect on the conditions-the microclimates-that prevail there and provide shade and physical structure for many other organisms that do not directly utilize the trees as a resource.

It should be clear that when we speak of ecosystem structure, we are considering various aspects of an ecosystem from a perspective that need not-indeed does not-mesh exactly with other ways of categorizing ecosystem components. Consider the trees of a forest or the coral heads of a coral reef. Each is a living organism, a resource used by many other species, and a physical structure that determines the uniqueness of its particular ecosystem compared with the environment that would occupy the same space in the absence of either trees or corals. In describing ecosystem structure, we do not break down the components into divisions of abiotic vs. biotic, inorganic and organic, or by species.

Functional Components

We can recognize four functional components of an ecosystem:

— abiotic substances,

— producers,

— consumers, and

— decomposers.

The latter three are living components, what Odum termed the three "functional kingdoms of nature".

Abiotic Substances

Abiotic substances are the inorganic and organic compounds of the environment. We studied many of these in detail in.

Producers

Producers are autotrophic organisms. "Autotrophic" means self-nourishing. The most conspicuous group of autotrophs are the photolithoautotrophs, organisms such as algae and flowering plants that have cells containing chlorophyll and are thus capable of fixing light energy ("photo-") to build complex organic substances from simple inorganic substances ("litho-").

Consumers

Consumers are heterotrophic organisms. These are also termed macroconsumers.

Decomposers

Decomposers are heterotrophic organsms. These are also termed microconsumers, saprobes, or saprophytes.

Biotic Community

A community or biocoenosis is an assemblage of organisms (species populations) whose composition and aspect is determined by the properties of the environment and by the relations of the organisms to each other. Möbius proposed the term biocoenosis in 1877 in his study of oyster reefs in the North Sea, recognizing that interactions between species in a structurally complex system brought with it a higher level of functional complexity.

Shelford defined a community as an "assemblage with unity of taxonomic composition and a relatively uniform appearance. The biotic community, along with the physical or abiotic elements of the environment together constitute the ecosystem of Tansley. Like "ecosystem", the size or area occupied by a community can vary considerably, bounded by the distribution of the component species populations where these overlap, producing a uniform species composition. Some species might, as well, be members of other surrounding (or distant) communities.

The community concept is broadly defined and one of nested units: conceptually small units potentially being part of containing larger units. This fact is recognized where ecologists define major communities as those of sufficient size and completeness of organization to be relatively independent of adjoining communities, and minor communities as those more or less dependent on neighboring (or surrounding) assemblages.

Communities show both functional and structural (taxonomic) unity, although some species may be replaceable over time and space so that functionally similar communities can have different species compositions. The community concept emphasizes the idea that organisms live together in an orderly manner; that species distributions are more than simply individuals responding in a similar way to a particular set of physical factors. And, of course, a community has properties or characteristics that are more than the sum of its species populations and unique attributes.

Ecological Balance

From molds to man, the entire living world lives off just a thousandth of all the energy that pours down to the earth from its central star, the Sun. This tiny fraction of the Solar energy is trapped by the green plants in the most massive of biochemical reactions, photosynthesis, and foods are elaborated. The-foods so built sustain the myriad variety of the living species.

Some of them straight draw their sustenance from the green plants so benefited by the Sun and others, indirectly. The earth itself supplies the water and the mineral requirements of the living world, contributing to the framework as well as the working of the body. In other words, the biological and the physical worlds from an inextricable integrated whole.

This understanding of Nature has led biologists to look for the inter-relationships that ramify through the different forms of life: the increase in numbers within one species has its own impact upon the numbers of other species. As long as Nature worked unmolested by man, proper balance among the constituents of the living world was maintained.

But when man began to bend Nature to satisfy his needs and, worse still, his fancy, he introduced an element of imbalance into its working. When forests yielded place to farms, it was not the landscape alone that suffered. The trees no doubt disappeared but with them also disappeared the deer, the tiger, the beetles, the birds, the lichens, the mosses, the mushrooms and a host of other forms of the living world.

Rodents and sparrows that feed upon the grains found a new haven in this new set-up. And that is not all. With the large trees also disappeared the great up-take of the soil water and its release into the atmosphere through transpiration. Consequently, rainfall suffered.

Subtle changes in the climate set in. Where in the beginning, the land yielded rich crop, over the years the farmer found to his chagrin his land was not as good a proposition as it used to be. He found it increasingly necessary to augment the water and mineral supplies. Pests that he did not suspect appeared and began to reap his harvest.

The tragic part of it was that it was all his own doing. Not that he was not aware of the great danger that was building up but he was too pre-occupied with his immediate gains to pay heed to the warnings. It took the disappearance of several species of animals and the near extinction of quite many others for him to wake up. The frightening increase in this depletion in the numbers of other species lent a new perspective. Ecology, the environmental science, acquired a new importance.

Understanding the mutual impact of the various species living in a given locality is also the subject-matter of ecological studies. It is hoped that with this understanding, there would grow a greater consciousness which would bring man and his technological civilisation in tune with Nature's schemes. The ecologist protests against the release of factor effluents into a nearby stream for it is not the fish alone that is affected but also the cattle and the men that drink this water and eat that fish.

At Malapadu in Andhra Pradesh, the cattle-heads suffered initially paralysis of limbs and finally death due to the lead-poisoning caused by the effluents led into a nearby stream from a factory that extracted copper from the rich copper ore of a neighbouring hillock. It is feared that men living in the area are in real danger of suffering a similar fate if they continued to draw their water supply from this stream.

The stately palmyrah palms of Ennore in Chennai have almost disappeared, thanks to the sulphurous fumes that keep pouring out of the chimney of the fertiliser factory there. So often, the effects are not this visible. They tend to be subtle and insidious. The ecologist warns us against these and his warnings can be ignored only at the peril of the human species.

Nature works in ecosystems. An ecosystem comprises a group of living organisms and their physico-chemical environment. A system is an entity in its own right. Disturbance at one point is sure to have repercussions all over the system and if not repaired on time, the system is liable to disintegrate. In every ecosystem, there is a balance naturally attained in which each component is in perfect harmony with the rest. When man clears a forest or builds a dam, he interferes with the ecological balance that has been there.

Without intending it, he disturbs the ecological balance of the area and is seeking to establish a new ecosystem. A new system can come only in the place of an old one. In the bargain, the component inhabitants of the old system disappear. When older ecosystems are upset over a large area, the migratory, the adaptive and such other limitations cause the total disappearance of a few species.

This is undesirable not only because the richness of the flora and fauna suffer but more so because in the new dispensation, there might be unforeseen and disastrous consequences. For instance, the rich American's fancy for money purses made of snakeskin let loose a ruthless rampage on rat snakes in south India.

2

Environmental Problems

Air Pollution

Air pollution is one of the most pervasive environmental problems because atmospheric currents can carry contaminated air to every part of the globe. Most air pollution comes from motor vehicle emissions and from power plants that burn coal and oil to produce energy for industrial and consumer use. Carbon dioxide and other harmful gases released into the air from these sources adversely affect weather patterns and the health of people, animals, and plants. It is estimated that three million people may die of air pollution each year worldwide. 2.8 million of the 3 million mortalities may be due to indoor air pollution. 90% of the 3 million estimated deaths are in developing nations. 70,000 die each year in the U.S..

Deaths from air pollution are compared to deaths from second hand smoke and chemical weapons. In the U.S, more people die from air pollution than from car accidents. They die specifically from agitated asthma, bronchitis, emphysema, lung and heart diseases, and other respiratory allergies. The EPA estimates that a proposed set of changes in diesel fuel technology could result in 12,000 fewer premature mortalities, 15,000 fewer heart attacks, 6,000 fewer emergency room visits by children with asthma, and 8,900 fewer respiratory-related hospital admissions each year in the US. The worst short-term civilian event from pollution in India was the 1984 Bhopal Disaster.

Leaked industrial vapors killed more than 2,000 people outright and injured anywhere from 150,000 to 600,000 others, some 6,000 of whom would later die from their injuries. The worst single incident of air pollution to occur in the United States of America occurred in Donora, Pennsylvania in late October, 1948, when 20 people died and over 7,000 were injured. The United Kingdom suffered its worst air pollution event when the December 4th Great Smog of 1952 formed over London. In six days more than 4,000 died, and 8,000 more died within the following months.

An accidental leak of anthrax spores from a biological warfare laboratory in the former USSR in 1979 near Sverdlovsk is believed to have been the cause of hundreds of civilian deaths. Intentional air pollution in combat is called chemical warfare. Poison gas as a chemical weapon was principally used during World War I, and resulted in an estimated 91,198 deaths and 1,205,655 injuries. Various treaties have sought to ban its further use. Non-lethal chemical weapons, such as tear gas and pepper spray, are widely used.

Sources

Anthropogenic Sources related to burning different kinds of fuel - human activity

— Combustion-fired power plants.

— Vehicles with internal combustion engines.

— Devices powered by Two-stroke cycle engines.

— Stoves and incinerators, especially coal ones.

— Wood fires, which usually burn inefficiently.

— Farmers burning their crop waste.

Other Anthropogenic Sources are

— Aerosol sprays and refrigeration, which once depended on Freon and other chlorofluorocarbons.

— Arsenic and chlorine found in drinking water and inhaled in bathroom showers.

— Dust and chemicals from farming, especially of erodible land.

— Fumes from paint, varnish, and other solvents.

— Industrial activity in general.

— Military actions, including the use and testing of nuclear bombs, poison gases, and germ warfare.

— Oil refining.

— Rocketry, which produces many tons of exotic emissions quickly and which deposits some of them directly into the tenuous upper atmosphere.

— Waste deposition in landfills, which generate methane.

Natural sources

— Dust from natural sources, usually large areas of land with little or no vegetation.

— Methane, emitted by the digestion of animals, usually cattle.

— Pine trees, which emit volatile organic compounds (VOCs) and oxygen.

— Radon gas from earth minerals.

— Smoke and carbon monoxide from wildfires.

— Volcanic activity, which produce sulfur, chlorine, and ash particulates.

Contaminants

Contaminants of air can be divided in particulates and gases. Particulates are small, solid particles, classified by their sizes. Atmospheric particles are usually measured as TSP, PM10 or PM2.5. TSP stands for Total Suspended Particulates. The PM10 fraction consists of particles with an aerodynamic diameter of less than 10 micrometres; these are more dangerous to humans than TSP, because they can be breathed deep into the respiratory tract and reach the lungs. PM2.5 particles are even more dangerous because they can pass through the upper airway filtering and into the alveoli, where they can cross the lung/blood stream barrier and transport into the blood.

Increasing attention is now focussing on the health impacts of even smaller particles- the so-called 'nanoparticles'. Smaller particles tend to be more toxic than larger particles and can stay airborne as an 'aerosol' for longer than larger particles, which settle out more quickly. Important pollutant gases include:

— Carbon monoxide, which is primarily emitted from combustion process, particularly from petrol vehicle exhausts due to incomplete combustion; the highest concentrations are generally found at roadside locations. Inhalation of high levels of carbon monoxide can cause headaches, fatigue and respiratory problems. According to the EPA (as presented in the 2002 World Almanac), 97,441 thousand short tons of carbon monoxide were released in the United States during the year 1999, 75,151 of those caused by transportation related exhaust.

— Chlorofluorocarbons, which destroy the stratospheric ozone layer.

— Hydrocarbons.

— Lead and heavy Metals

— Nitrogen oxides, or NO_x. Emissions are primarily in the form of NO, which is oxidised by ozone (O_3) from nitric oxide to NO_2. Nitrogen dioxide (NO_2) is the primary concern for effects on health, and is the species for which WHO health-based standards are expressed. The various oxides of nitrogen can also react with hydrocarbons in the atmosphere to contribute to photochemical smog. NO_x can also affect ecologically sensitive sites through deposition, causing acidification and eutrophication. In The U.S., 25,393 short tons of Nitrogen Oxide were released during 1999.

— Sulfur oxides, which causes acid rain is caused from the burning of fuel containing sulfur, mostly at power plants, and during metal smelting and other industrial processes. In the U.S., 12.46 tons of sulfur dioxide were released in 1999, however there has been a 33 percent decrease in emissions between 1983 and 2002, due largely to state restrictions.

— Tropospheric ozone, which is ozone in the lower part of the atmosphere. Ozone (O_3) is a secondary pollutant, formed through photochemical reactions involving NO_x and hydrocarbons; it is an irritant gas. In the stratosphere it helps to reduce the amount of ultraviolet radiation from the sun that reaches earth).

— Volatile organic compounds: gasoline, solvents, cleaning solutions.

Indoor Air Pollution

The lack of ventilation indoors concentrates air pollution where people are most exposed to them. Background pollution comes from such mundane sources as shower water mist containing arsenic, which is damaging to inhale. The arsenic can be trapped with a shower nozzle filter. Radon gas, a carcinogen, is exuded from the earth and trapped inside houses. Researchers have found that radon gas is responsible for over 1,800 deaths annually in the United Kingdom. These natural radon emissions can be blocked by a layer of aluminum foil under the carpet (according to the U.S. Department of Air Quality Management).

Building materials including carpeting and plywood emit formaldehyde gas. Paint and solvents give off volatile organic compounds (VOCs) as they dry. Lead paint can degenerate into dust and be inhaled. Asbestos insulation was commonly used in many application and can be carcinogenic in the lungs. Intentional air pollution is introduced with the use of air fresheners, incense, and other scented items. Controlled wood fires in stoves and fireplaces can add significant amounts of smoke particulates into the air, inside and out. Clothing emits perchloroethylene for days after dry cleaning.

Deaths are often caused by using pesticides and other chemical sprays indoors without proper ventilation, and many homes have been destroyed by accidental pesticide explosions. Second-hand tobacco smoke is now recognized as an indoor air pollutant which accounts for an estimated 3,000 lung cancer deaths annually in the US. Carbon monoxide poisoning is a quick and silent killer, often caused by faulty vents and chimneys, or by the burning charcoal indoors. 56,000 Americans died from CO in the period 1979-1988.

Chronic carbon monoxide poisoning can result even from poorly adjusted pilot lights. Smoke inhalation is a common cause of death in victims of house fires. Traps are built into all domestic plumbing to keep deadly sewer gas, hydrogen sulfide, out of interiors. Biological sources of air pollution can also be found indoors, and include gases, particulates, allergens, and microbes. Pets produce dander, bed mites deposit shells and microscopic droppings, inhabitants emit methane, mold can form in walls and generate spores, air conditioning systems can incubate Legionnaires disease, toilets can emit feces-tainted mists, and houseplants and surrounding gardens can produce pollen, dust, and mold spores.

Water Pollution

Water pollution has many sources and characteristics. Humans and other organisms produce bodily wastes which enter rivers, lakes, oceans and other surface waters; in high concentrations these wastes result in bacterial contamination and excessive nutrient loading (eutrophication). Industries discharge a variety of toxic compounds and heavy metals, and wastewater from industrial process may also be too hot or too low in dissolved oxygen to support life. Silt-bearing runoff from construction sites and farms can inhibit the penetration of sunlight through the water column, hampering water organisms in their ability to photosynthesize.

Groundwater pollution is more difficult to clean up than surface pollution because groundwater can move hundreds of miles through unseen aquifers. Porous, fine-grained aquifers such as sands and sandstones naturally purify water of bacteria by simple filtration, but have no effect on chemical contaminants such as the components of gasoline. Groundwater that moves through cracks and caverns is not filtered and can be polluted just as easily as surface water. In fact this can be aggravated by the human tendency to use sinkholes in areas of Karst topography as dumps.

Causes

The causes of water pollution can be divided into two groups: anthropogenic sources of pollution are those due to human choices, and natural sources are those resulting from forces intrinsic to the environment.

Anthropogenic sources include:

— discharge of poorly-treated or untreated sewage;

— runoff from construction sites, farms, or paved and other impervious surfaces e.g. silt

— discharge of contaminated and/or heated water used for industrial processes

— acid rain caused by industrial discharge of sulfur dioxide (by burning high-sulfur fossil fuels)

— excess nutrients added by runoff containing large amounts of detergents or fertilizers

Natural sources include:

— seasonal turnover of lakes and embayments;

— siltation due to floods;

— eutrophication of lakes due to seasonal changes

— acid rain caused by natural volcanic discharges or smog from factories

— acid pollution of rivers and lakes by runoff from naturally acid soils

— carbon dioxide discharges and runoff, volcanic or mineral

Contaminants

Contaminants may include organic and inorganic substances. Organic water pollutants include:

— bacteria, as from sewage or livestock operations;

— fertilizers, in runoff from agricultural fields or forestry;

— food processing waste;

— tree and brush debris from logging operations

Inorganic pollutants include:

— metals

— acid mine drainage

— acid rain caused by industrial or volcanic discharges

— acid pollution of lakes by runoff from acid soils

— carbon dioxide discharges and runoff, volcanic or mineral

— chemical waste industrial byproducts

— silt in stormwater runoff from cleared land

Consequences of Water Pollution

The toxic chemicals and particles are carried out by the rivers into the ocean. In some areas of the world the influence can be traced hundred miles from the mouth, like in front of the Hudson river. As indicator filter feeding animals are used by the oceanographers, like copepods in the map of New York Bight. The highest toxin loads are not directly in front off New York but 100 km South, because it takes a few days to be incorporated in the tissue of the plankton.

The Hudson water flows south along the coast due to the coriolis force. The second map shows areas of oxygen depletion, caused by chemicals using up oxygen and by heavy algae blooms, caused by too much nutrients, when the cells die, sink and decompose. Heavy fish and shellfish kills have been reported. The toxins make their way up the foodchain when small fish eat the copepods, then large fish, each step concentrating up ca. 10 times.

Threats to Biodiversity

A diverse ecosystem is important. Biodiversity actually

boosts ecosystem productivity where each species, no matter how small, all have an important role to play. It is this combination that enables the ecosystem to possess the ability to prevent and recover from a variety of disasters. This is obviously useful for mankind as a larger number of plant species means a greater variety of crops. In addition, a larger number of species of animals ensures that the ecosystem is naturally sustained. And so, while we dominate this planet, we still need to preserve the diversity in wildlife.

At least 40 per cent of the world's economy and 80 per cent of the needs of the poor are derived from biological resources. In addition, the richer the diversity of life, the greater the opportunity for medical discoveries, economic development, and adaptive responses to such new challenges as climate change. A healthy biodiversity provides a number of natural services for everyone:

1. Ecosystem services, such as
   — Protection of water resources
   — Soils formation and protection
   — Nutrient storage and recycling
   — Pollution breakdown and absorption
   — Contribution to climate stability
   — Maintenance of ecosystems
   — Recovery from unpredictable events
2. Biological resources, such as
   — Food
   — Medicinal resources and pharmaceutical drugs
   — Wood products
   — Ornamental plants

— Breeding stocks, population reservoirs

— Future resources

— Diversity in genes, species and ecosystems

— Social benefits, such as

— Research, education and monitoring

— Recreation and tourism

— Cultural values

That is quite a lot of services we get for free! Now, as we erode ecosystems and reduce biodiversity the cost of replacing these (if possible) would be extremely expensive. It therefore makes economic and development sense to move towards sustainability. While there might be "survival of the fittest" within a given species, each species depends on the services provided by other species to ensure survival. It is a type of cooperation based on mutual survival and is often what a "balanced ecosystem" refers to.

As well as the need for biodiversity for continued ecosystem survival, from a human perspective, most of our medicine (from common drugs to possible cures for cancers) come from plants, many of which are now endangered. However, it is feared that human activity is causing massive extinctions. A huge report known as the Millenium Ecosystem Assessment, started in 2000, was released in March 2005. Amongst many warnings for humankind, it noted that there has been a substantial and largely irreversible loss in the diversity of life on Earth, with some 10-30% of the mammal, bird and amphibian species currently threatened with extinction, all due to human actions.

Research of long term trends in the fossil record suggests that natural speed limits constrain how quickly biodiversity can rebound after waves of extinction. Hence, the rapid extinction rates mean that it could take a long time for nature to recover.

Loss of Forests

A 20-year study has shown that deforestation and introduction of non-native species has led to about 12.5% of the world's plant species to become critically rare. (In fact, as an example, a study suggests that the Amazon damage is worse than previously thought, due to previously undetected types of selective logging and deforestation.) A report from the World Commission on Forests and Sustainable Development suggests that the forests of the world have been exploited to the point of crisis and that major changes in global forest management strategies would be needed to avoid the devastation.

What also makes this a problem is that many of the endangered species are only found in small areas of land, often within the borders of a single country. New species of animals and plants are still being discovered. In Papua New Guinea, an 44 new species of animals were discovered recently in the forests. Logging may affect these animals' habitats, though.

The loss of rainforests around the world, where many species of life are found will mean that potential knowledge, whether medicinal, sustenance sources, or evolutionary and scientific information etc. could be lost. Brazil which is estimated to have around 55,000 species of flora, amounting to some 22% of the world's total and India for example, which has about 46,000 and some 81,000 animal species, are also under various pressures, from corporate globalization, deforrestation, etc. So too are many other biodiverse regions, such as Indonesia, parts of Africa, and other tropical regions.

Sustainable Forests

The overly corporate-led form of globalization that we see today also affects how natural resources are used and what priorities they are used for.

Some government institutions even buy illegal timber from pristine forests. For example, it is claimed that UK buys all of its Mahogany from pristine forests in Brazil where 80% of all timber is traded illegally. Even though Brazil has now tried to introduce a moratorium on Mahogany logging for two years, this has been slammed by some as too little, too late. Under much secrecy, there is a push from USA and Asian economies to reduce tariffs for wood and paper products. Also at the WTO Ministerial meeting in November 1999, opening more markets for easier access was the agenda, which included forests.

People and Forests

Quite often we make blanket statements - or worse still, generalized conclusions - along the lines that people are the cause of deforestation. Of course that is true. However, unfortunately all people around the world are not equal, and it also also follows that some are more responsible for deforestation than others. Often, in forests of the Amazon, Africa, or Asia, forest protection schemes have been promoted that go against indigenous peoples and cultures, rather than work with them.

As Indian activist and scientist Vandana Shiva and others have shown in countless work, indigenous people often have their cultures and lifestyle structured in a way that works with nature and would not undermine their own resource base. For example, in her book Stolen Harvests she describes how their traditional knowledge has been beneficial to the environment and has been developed and geared towards this understanding and respect of the ecosystems around them. Yet because of blanket conclusions that humankind is responsible for deforestation, we risk assuming all types of societies are equally responsible for deforestation that is damaging to the environment.

Global Threats to Coral Reefs

All around the world, much of the world's marine biodiversity face threats from activities and events such as

— Coastal development;

— Overfishing;

— Inland pollution;

— Global climate change.

Global change threats

These are:

— Coral bleaching - caused by elevated sea surface temperatures due to global climate change;

— Rising levels of CO_2

— Diseases, Plagues and Invasives - linked to human disturbances in the environment.

Direct human pressures

These are:

— Over-fishing (and global market pressures) - including the use of damaging practices (bomb and cyanide fishing);

— Sediments - from poor land use, deforestation, and dredging;

— Nutrients and Chemical pollution

— Development of coastal areas - for urban, industrial, transport and tourism developments, including reclamation and mining of coral reef rock and sand beyond sustainable limits.

The human dimension

These are:

— Rising poverty, increasing populations, alienation from the land

— Poor capacity for management and lack of resources

— Lack of Political Will, and Oceans Governance

Impacts of Climate Change

The major emerging threat to coral reefs in the last decade has been coral bleaching and mortality associated with global climate change. It is believed that almost all species of corals were affected by high sea surface temperatures during 1998 which led to global coral bleaching and mortality. 2002 has been the second worst year for coral bleaching after 1998. Scientists seem to be pessimistic about the future, with some reefs expected to vanish by 2020.

Additional scientific research, reported by Greenpeace fears climate change will eliminate reefs from many areas: If climate change is not stopped, coral bleaching is set to steadily increase in frequency and intensity all over the world until it occurs annually by 2030 - 2070. This would devastate coral reefs globally to such an extent that they could be eliminated from most areas of the world by 2100. Current estimates suggest that reefs could take hundreds of years to recover. The loss of these fragile ecosystems would cost billions of dollars in lost revenue from tourism and fishing industries, as well as damage to coastal regions that are currently protected by the coral reefs that line most tropical coastlines.

Legacy of Nuclear Tests

In 1995, France started testing it's Nuclear weapons in the Pacific despite huge protests. It is now emerging that the coral in the French Polynesia regions where many Nuclear tests have been carried out have been harmed, as the French atomic energy commission has admitted. This is raising concern over what else they may have failed to tell the people who have to live through it in that area.

Pollution of waters and oceans is increasing. It is easy to take clean water and regular access for granted in the developed nations. However, most of the planet's population cannot get clean water.

While it has been shown, and is widely accepted that many governments have done a poor job at providing basic services, such as water access, often supported solutions seem to be quite extreme, such as privatizing water. They also describe the inequality of water consumption and attacks the issue of privatizing water as risking commodification of a basic necessity and a public service, which "reduces the involvement of citizens in water management decisions". On the one hand, as environmentalists often point out, the value of water has been underestimated and hence overused.

There is a practical problem in some areas of getting appropriate incentives to improve and enhance water-related systems and access. Private companies have resources that could be mobilized on this, but criticisms have been abundant regarding that the conflicts of interest between shareholders and the need for profit, versus universal access to water has not worked. Indeed, in some places such as Bolivia and elsewhere, where the nation has had to sell of state-owned water companies to private corporations price hikes have resulted and led to civil unrest and protest. Similar problems in other places countries around the world, such as Panama, Ecuador,

Argentina, South Africa and elsewhere have not escaped mainstream media attention, either.

Water and the Ecology

The deterioration of ocean ecosystems can also affect the whole planet. For example, while scientists claim that it is very difficult to predict the effects of Global Warming, some believe that it could cause Europe to become much colder than it is at present. A report from WWF finds that marine life patterns are changing due to rising global temperatures. It is disrupting a range of species in the food chain from planktons to polar bears. Many of the species affected are already facing decreasing numbers. This could affect humans too as we rely on the oceans for fishing, tourism etc.

Access to fresh water seems as if it is beccming a political problem, rather than a technical one with lots of questions on the best way for countries to provide it. In Uruguay, a system for improving access to water had been developed and introduced in 1998 which hoped to increase the availability of drinking water to over 1.2 billion people around the world who do not have regular access. Due to the fairly small investments, very low operation costs and quick installations, it was hoped that this will aid a practical solution towards increasing shortages around the world. However, not much seems to have happened since.

Some activists were concerned about the corporate agenda in water privatization. However, as per the final declaration of the water forum, water security was defined to mean that "freshwater, coastal and related ecosystems are protected and improved; that sustainable development and political stability are promoted, that every person has access to enough safe water at an affordable cost to lead a healthy and productive life and that the vulnerable are protected from the risks of water-related hazards".

POPULATION

Many feel that the major international wars to be fought in the future will continue to be over natural resources. Power conflicts and self-interest will perhaps mean that there will be gross violation of basic rights and death or misery for millions of innocent people. Throughout history, most wars have had trade and resources at their core (leading to ideological battles) fueled by imperialistic motives. In the future, while this pattern is likely to continue, as resources get depleted and wasted in these wars (hot and cold), additional conflicts and contention will arise through access to even more limited resources.

Many of us have grown up learning and being told that 6 billion is too much and this "over population" is primarily impacting the planet's ability to cope. But is that really the case? Sure, the planet is facing incredible stress. But how much of that is due to large populations, and how much is based on other factors, such as how we choose to live, how we produce, consume and waste our resources? The poor are numerous, but as we shall see, consume far less resources of the planet, for example. Studies point to ecological limits to sustain people, but these limits can be different, based on the way we consume resources etc so it is hard to say for sure what over population means let alone if we are at some threshold, below, or above it.

The information understood so far provides valuable insights and is very important to consider, nonetheless. Yet, the figure of 6 billion and literature about over-population naturally looks to the poor regions where there are high populations and environmental degradation as the problem. Do large populations affect and put stress on the environment, society and resources? Populations do affect and put stress on the environment. However, some claim that overpopulation is the major cause of environmental degradation. Is that so?

While populations no doubt are large in many countries, and demands on resources are obviously large, it is only one of many other causes and some of those other issues such as over-consumption based, unsustainable development may have an even larger impact. Our choice of how to use those resources (i.e. our economic policies) and for what purposes (i.e. our political directions and policies) are critical issues as well on the resulting impact on the environment to meet those uses and purposes.

Consumption

Existing consumption patterns as seen in Europe and North America can put strain on the environment and natural resources, which can have serious impacts on society. But, how much of the environmental degradation we see today is as a result of over-population and how much is due to over-exploitation due to consumerism and geopolitical interests?

Some suggest that the industrialized nations need to drastically change their consumption patterns that are currently seen, as this is depleting resources more than the demands from large populations as seen in many developing nations. Consumption patterns driven by current growth-based economics promote production and purchase of more and more material wealth and satisfaction. The impact on the environment from the increased need to obtain inputs such as raw materials and use of the environment, as well as increased waste, is considerable. The consumption levels of all people around the world are not the same.

If countries with large populations such as China and India follow the consumption models of today's wealthy nations, then the impact on the planet will be considerable. And China for example, is interested in increasing the use of automobiles. Automotive business interests are mouthwatering at the prospects of a massive "market" to build and sell cars, regardless of concerns on the environment. In addition, if we look at consumption disparities today, as shown above, it also helps put another perspective on the issue of population problems and consumption problems with respect to strains on the environment.

Furthermore, if we conclude (as some resign to) that changing economic models and consumption habits are far too difficult, is the alternative to reduce population sizes in the poor countries? What of the fact as shown by the UNDP report quoted above, that the impacts and demands on the environment by the poor is far less when compared to the rich? If a sustainable planet is still the aims and goals, then in admittedly simplistic terms, one would have to do some 86 times the effort to get the same supposed gains by reducing poor populations! Reducing populations in poorer countries without addressing economic justice issues and so on could lead to further poverty and marginalization of other people while giving a rest bit to the environment, further legitimizing calls that the poor are to blame for most environmental degradation.

Misuse of Resources

How land is used to produce food etc. can have enormous impacts on the environment and its sustainability. Because industrial agriculture is using more monocultures, rather than a diversity of crops, the loss of biodiversity is leading to more resource usage, as described above. This as well as other political situations such as the motives for dumping surplus food on to developing countries to undersell the local farmers, leads to further hunger around the world. Consumption patterns in wealthier countries increase demands for various foods, flowers, textiles, coffee, etc.

Combined with commercial interests in things like tobacco, largely grown by corporations from wealthy nations, and with input-intensive agricultural practices (including using herbicides, pesticides and fertilizers, even if their use is becoming more technically efficient) the diversion, and misuse of land and the associated environmental damage in unsustainable methods adds up. Land ownership has become more concentrated in the hands of larger companies, larger agribusinesses and so on. In addition, things like food dumping, mentioned above, increases hunger and drives rural workers out of jobs.

Such effects combine and lead to an increase in urban migration as people move to the cities in hope for a better chance. These economic policies that are based less on people's sharing and development, but more on acquiring wealth and profit lead to additional stress on the larger cities to provide for more people. It also results in more slum areas, health problems and so on. Many easily conclude that by just looking at the cities that we have overpopulation in the world. While the cities are no doubt facing problems of over population, a variety of political and economic circumstances are leading to such conditions and looking only at cities to determine if the planet is over populated misses out these factors.

But cities aren't the only places that the landless move to. Some, being pushed off their own lands, will move to less arable land and forest frontiers in the hope to farm that, which may conflict with wildlife. In other cases, many may try to immigrate to other parts of the world if they feel there is no choice left in their own country. In yet other situations, economic growth can also lead to more urban migration. Sometimes this growth of cities can go in hand with decline in the rural areas.

Due to these and a multitude of other complex socioeconomic and political factors, in different parts of the world, there are different proportions of people in urban and rural areas. For example, the World Bank reported in a 1999/2000 report that 74% of poor in Latin America and Caribbean lived in urban areas, while in Europe and Central Asia it was 67%. In the Middle East and North Africa it was 58%. In East Asia and Pacific, 33% while in Sub-Saharan Africa it was 32%. In South Asia it was 27%. North and Central America have approximately 76% and 50% urban populations, respectively.

Economic policies of the wealthier nations and their consumption demands mean that more land is often used for less than ideal purposes, such as

— Growing cash crops (bananas, sugar, coffee, tea etc) for export to wealthier countries (primarily);

— Diverting productive land for non-productive uses (tobacco, growing flowers for export markets, etc);

— Clearing land and used to grow things like cattle for beef exports. (Parts of the Amazon for example, are cleared for cattle ranches so the beef can be exported for use in fast food restaurants. Smaller parts are cleared by the poor who are forced onto marginal lands, or frontier areas due to poverty.

These economic policies are often imposed on the poorer nations, through things like Structural Adjustment (SAPs). In the past, colonialism achieved similar things more explicitly, but today it is often wrapped up in complex trade and economic agreements.

Unsustainable Development

In many regions of the world, ecosystem stress can be seen due to human-intensive activities such as unsustainable resource extraction and exploitation via large corporations and international trading agreements that open these sources up for excessive extraction and consumerism. This could lead to numerous future global security issues. At the same time, many businesses and individuals try to calm fears claiming that nothing is wrong. Major themes used for this are:

— To remind us of mans ingenuity to deal with any problems that may occur;

— That the resources are abundant enough to not cause a problem and even if they do, we will adapt;

— And that history shows that we have always come out for the better.

This cornucopian view that everything is ok, has come under much criticism, using its credible data in a partial way. It is also seen to be an excuse to avoid responsibility and therefore welcomed by some corporations who may somehow need reassurance or justification if they are using some exploitative economic measure. Some parts of the criticism itself has come under scrutiny because of anti-immigrant and racist sentiments in some of the environmentalists who argue against the cornucopian views and is therefore part of a continuing debate on the causes and effects of population growth and if current population growths are even a problem.

It has been noted however, that in developing countries large cities are increasing in population at a rate that is very difficult to sustain and provide for without better economic growth. This could lead to various social problems such as an increase in hunger and crime. However, note that there is often a distinction between population explosions or growth, and cities increasing in population density. Economic and other social and political conditions may lead to many people from rural areas moving to the larger cities in search for jobs. If unmanaged, or root causes for these movements are not addressed, this can place a burden on the ability for a large city to provide in a sustainable manner.

Poverty can lead to environmental degradation, no doubt. It can also lead to a decreasing availability of land and water in some areas. Having then suggested it is not always large populations that are ultimately the main cause of environmental stress etc, larger populations, global warming, industrial pollution, high-tech agriculture, even misplaced development priorities as well as over-consumption and non-sustainable development can all put increased pressure on the environment to provide the resources such as enough arable land and fresh water.

But to say it is only "over" population as the root cause of environmental problems of this nature is not enough. It is one dynamic of this complex issue. An over-consumption based approach to development may even be a larger, more potent factor. Hence, sustainable development becomes even more important overall. Better management of natural resources, improved education in the area of reproduction-related issues as well as fairer economic and social and political environments etc., would help ensure "prophesies of doom" are averted.

Ecological Footprints

For some cities, or even countries, the ecological footprint can be even larger than that nation. For example, Britain's footprint has covered many countries, especially during the imperial days. And today, London's ecological footprint for food, forest products etc has been estimated to be 120 times the surface area of the city itself, nearly equivalent to the productive land area of Great Britain as a whole. But it isn't just cities from industrialized nations that have large ecological footprints; even growing cities in developing nations, such as Delhi, are seeing their footprints increase, and with it, the number of associated problems are also rising.

With the above U.N. statistic that the world's wealthiest 20 percent of humanity consume 86 percent of the world's resources, we can roughly say that the ecological foot print of the wealthier people is pretty much the rest of the planet. So, for those wondering why the poor cannot follow the example of the rich and get out of poverty themselves, here lies the answer: There is nowhere from which to get out of poverty. If resources are attempted to be reclaimed, it is a threat to those who currently use it. Wars throughout history have often been related to this. World War II and the resulting Cold War were such battles.

Some people have researched the carrying capacity of various areas or even the planet and have concluded that the planet can only sustain a certain amount of people. However, those estimates range from some 6 or 7 billion to even a few hundred billion people. How can these carrying capacity estimates numbers vary so much? A lot depends on the various assumptions which are invariably made to get these numbers. One thing often overlooked is the fact that as economic circumstances change, consumption habits and patterns will change. Therefore, the demand on the planet for sustaining us also changes, as does the carrying capacity.

The "Science" of Ecological Limits

The science of ecological limits is often used to conclude that population growth and numbers are the major cause of environmental degradation. However, while the study of ecological capacities and limits is important, some of the conclusions are a major source of contention. Some conclude, looking at the usually credible data of the effects, that the causes of environmental degradation are due to populations by virtue of numbers and that as a result, we are "over" populated. Yet, these are not automatic conclusions and cannot be taken as a given. Such conclusions are even flawed due to the number of assumptions made as well as the number of other important issues not taken into account.

Many environmentalists and others who study ecological problems and conclude population pressures to be the root, point out quite accurately and with detailed knowledge, the various limits on different ecologies based on current patterns of human activities, which they also show are being reached or even breached. At a very high level, this is true. Humans require resources to survive. Those resources come from the environment. And hence, non-natural environmental degradation is a result of human activities.

However, from this high vantage point, where all people and their impacts on the planet are seen as "equal", inappropriate conclusions form on where the root problems come from. Hence, "over" population (in the poorer countries) is the often-attributed root cause. However:

— Population/demographic issues are just one aspect of "human activities."

— How we organize ourselves to make use of resources - i.e. our political and economic choices - is another major important aspect.

— So too is for what reason we make use of resources, as it is not an obvious assumption that it is to meet all our needs.

— Also important is how all these things interact with each other, and with the many other facets of human activities. For example, economic/political causes of population growth (and decline) are often not factored in, resulting in assumptions based on religious beliefs, uncontrollable urges, lack of contraceptives, and so on.

— By making this base assumption of looking at population issues as a root cause, rather than a symptom of others, because that is so obviously a human impact, the risks of promoting ineffective policies, and even blaming victims of deeper causes increase, while underlying causes remain.

— Furthermore, changes in these root causes would change the limits of ecologies to sustain a different lifestyle. These major "variables" would lead to different answers in these complex equations.

Observing symptoms such as environmental degradation, hunger, disease, population growth, poverty and so forth, the causes are assumed to lie in population growths. After all, the ideas, while simple, are appealing and make sense at a scientific and basic economic level; that as population grows, demands on the environment increase, demands on food increase etc and if not kept up, more people will go hungry, more people will be poor etc. Hence, from the enormous, often very credible scientific data on impacts, attempts at concluding the causes of those impacts are affected by the range of assumptions and even cultural discourse that can define those ranges of assumptions indirectly or directly.

That is, factors, as mentioned above, such as our choice of not only what agricultural methods we follow, what foods we grow, what lands are cleared and used etc, but also how these things are done, can all be affected by different drivers that determining the purpose of these activities. It is easy to assume that these activities of things like growing food, clearing land etc is to meet human population demands of basic needs and that as populations grow, that these demands increase and therefore mean an increasing demand on the environment to provide those resources.

Agricultural decisions, economic drivers, political, geopolitical and social/cultural influences are all enormously complex variables in the equations of ecological limits, yet, many scientists fail to look at these influences on things like environmental degradation. Instead, more readily observable tangibles, such as large population sizes where there is poverty, hunger, disease, environmental degradation, etc are used as "proof" that these symptoms result from growing populations and the demands they place on the environment.

Critics will point out that the poor want to consume like the rich.

— Right. And such consumption will increase problems no doubt.

— Larger population numbers consuming in these wasteful ways will add to the problem.

— Hence, it is important to address these models of consumption that are now becoming global (via globalization). For example, the spread of the so-called "MTV culture" to other cultures are showing people what they could consume. This helps in creating demands and fashions (as fashion-driven consumption means even more use of resources, and wasting of existing resources etc).

— It is not that the poor should be denied increased consumptions but more that the form of consumption being promoted is what is causing the most environmental degradation and it is that form that is being spread globally.

— Furthermore, as will be seen below, the poor can improve their lifestyle, while also addressing inherent waste structured into the current systems of resource usage, i.e. politico economic systems.

— Additionally, this "attack" that is also made on immigrants is a double blow for them:

a) On the one hand, they are sometimes given incentives to come to the wealthier countries to help in the economy, or, they may choose to escape situations from their regions (sometimes related to global policies enforced or strongly pushed by those same wealthier countries), etc.

b) Yet

- If they try to remain a close group that "stick to their own kind" then they are criticized for not wanting to integrate with the society, of therefore being a cause of tension and so on.
- On the other hand, when they do attempt some form of integration (such as consuming the way the society in which they now live consumes) then they are blamed for being a burden on environmental resources and that immigration should be controlled or stopped because it is a cause of environmental degradation!
- It is almost like saying (whether intentional or not) "we don't want the poor to have a comfortable lifestyle like us. Too many poor having our lifestyles will mean less for us and the environment!"

— Most of the world's resources are not being consumed for the world's majority of people. Hence, it is difficult to say honestly that the world's poor and largely populated countries are the major cause of environmental degradation.

— As a result, the point that the poor want to consume like the rich indicates that it might not be the consumption of the poor at issue, but the consumption of the rich.

As a result, depending on how one looks at it, population issues are not easily reduced and understood by science alone, but the impact of many issues in combination. Scientific methods and research can provide much needed understanding on the impacts, but risks false or partial conclusions on the causes. This can be because such ecological science does not factor in politics, economics etc and almost (intentionally or unintentionally) assumes these things to be constant, rather than extremely variable.

By combining a variety of fields of study, it is more likely that a better understanding of causes will result. It is not that ecological sciences on these things are useless. They are not. As mentioned above, they provide useful statistics, data, trends etc. Predictions of future trends and as well as understanding of causes leading to the observed statistics, data and trends require a combination of many other fields of study and expertise. Furthermore, as a result, the causes of population growth and declines are often also missed out.

Solutions such as family planning, on their own do not consider factors about how populations grow. Instead, they address what is commonly seen as ecological degradation, where it is assumed that it is due to population growth, without understanding why populations grow. Poverty often results in other symptoms of such as widespread disease and outbreaks that could have been dealt with had more been spent on health. It is not population burdens that reduce health expenditure per person.

For example, structural adjustment policies from the wealthy demand that the poor cut back on health expenditure, which means even health education that promotes prevention is reduced or eliminated (and is cheaper than reactionary and curative approaches as that also expends a lot of resources). Yet, when cures for outbreaks etc are actually needed, multinational pharmaceutical companies who spend billions on producing drugs promote the view that governments should not interfere with their "freedom" to pursue research (for profits!) etc, produce drugs and cures for people who can pay. Instead, they often concentrate on less important things like baldness cures, anti-aging remedies etc while millions die from easily preventable diseases in poor areas. This is because, while the poor are a market, they cannot pay.

Another fundamental issue is that food and health are increasingly becoming commodities. That is, it is the ability for people to pay for these and other things that govern whether or not they will go hungry, ill etc. Hence it is in the root causes of poverty that determine who gets to consume resources, and as a result, what affects environmental degradation and so on. Economic policies of things like food "aid" (i.e. food dumping) are known and shown to increase world hunger, even though food production so far has more than kept up with the pace of population growth.

At the same time, international political/economic "agreements" result in using this precious land for things like tobacco, tea, coffee, floriculture, wasteful damn projects, sugar, beef and so on. Some of these things, such as sugar and beef are "luxuries turned into necessities" as not all consumption demands are natural demands; some demands and habits can be promoted and encouraged over others. Hence those who are more wealthy (and therefore more powerful) either economically, politically, or both, can be more influential in affecting the uses of our resources.

As an aside, vegetarianism therefore has another meaning if you think about it this way!

— Vegetarianism (or a large reduction in meat consumption) indirectly would help free up land for other uses such as growing food for others to eat as well - or in the case of beef consumption, help in reduce the pressures on natural forests such as the Amazon.

— Vegetarianism (or a large reduction of meat consumption) in an indirect way, could be a choice for those wishing to play a part in helping combat world hunger, environmental degradation etc.

— Likewise, reducing or eliminating tobacco and alcohol consumption can also be seen as indirectly helping address world hunger and environmental issues.

— Tobacco for example, is very water and nutrient-thirsty, hence less tobacco demand in theory would help stave off some environmental degradation if positive alternatives are appropriately supported, both politically and economically.

— If demands for such products were to decrease, that land could be used to grow other things such as food to feed the local people.

— All the support industries to promote, market and sell the consumption of such products, is, paraphrasing J.W. Smith's book title, "wasted wealth" by what he describes as wasted labor due to wasted capital.

— Of course, these alternatives cannot work in isolation. Economic alternatives also need to be addressed for the farmers and others who would lose out. This hints towards the need to address systemic and deep changes at the core, but this could be a starting point for people to research into issues of causes of world hunger, poverty, inequality, of additional anti-tobacco campaign themes and so on!

Of course, it is more complex than that, as political aspects of land control and its use still need to be addressed. Sugar, as another example, has historically required slavery-based plantations, and even today, sugar is grown in vast amounts, especially in various poor nations, for consumption of fast foods, soda drinks, candies/chocolates sweets and other confectioneries. Indirectly, there are additional enormous environmental costs that are not usually factored in. That is, the costs of using resources in the industries that support these things. As a very incomplete example, many resources are required:

— To create, maintain and support the office buildings where people work in these industries

— To support the marketing

— To support efforts in creating demands as well as meeting real and resulting demands

— To distribute and sell

— To create new ideas and products

— To create, maintain and support factories to make the actual products

— To create the materials for packaging

— To deal with the waste/disposal of these packages

— To deal with resulting health problems and the resources used to deal with them

— To pay and support lobbyists to help governments and regulation agencies see their perspectives and so on.

An entire industry is a "wasteful industry" when looked from a social and ecological perspective - but an efficient one when looked at from profit and capital accumulation perspective. We can see then that there are enormous impacts on the environment from things like:

— The politics of globalization (in its current form) is shown to increase poverty;

— All the effects of these policies including wasteful consumption contributes to further environment and resource usage

— Additional resources employed to support such a system also contributes to wasted resources

There is sometimes support for a simplified but somewhat flawed Malthusian view that population issues will cause hunger and poverty. However, other factors and assumptions are missing, such as:

— Why populations grow and decline

— How land gets used and what the impact of concentrated ownership is

— How the politics of economics affect societies and the environment

— What the effects of an economic system geared toward the use of resources for the purpose of accumulation of capital (which is a cyclic pattern seen throughout history) rather than to use resources efficiently to meet societies needs.

— And so on.

— These are all important "variables" that require more attention.

And what should be considered, as mentioned above, is the U.N statistic that 86% of the world's resources are consumed by the world's richest fifth of humanity, while the world's poorest fifth consume a "miniscule" 1.3% of the world's resources. This reflects the political/economic policies by the world's more powerful nations (and hence more wealthy) and the imbalance it causes. Resource usage, whether to meet basic human consumption needs, or materialistic consumption needs etc affect the environment. The phrase "over population" is a qualifying, or measuring, phrase due to the word "over".

That qualification has to be in relation to the environment and how we use its resources (as people live by using resources around them, which come from their environment). Hence, looking at how resources are consumed and who consumes what, how efficiently and so on would indicate the major causes of environmental degradations. We must look at these areas because it is not a given that resources are consumed to meet basic needs for everyone. When looking deeper then, the world's majority is not the major cause for the world's environmental destruction and stress on the planet. As a result, while we might be "over" something, is it population or consumption?

Humans do affect and put stress on the environment. However, the way they do is not necessarily a simple result of "over" population. Population related burdens on cities and resources are an issue, nonetheless, especially how they are consumed. It is only one of many other causes and some of those issues such as over-consumption based unsustainable development may have a larger impact on the environment. Because this latter process largely lies in the domain of industrialized countries, it is easier instead to blame the "others" who apparently "cannot control their population growths." It allows current ideologies of neoliberalism to continue without discourse of the ideology itself, just the symptoms.

As these neoliberal policies are also introduced in poorer nations, the wealthy in those countries have seen their wealth increase even more, while the poor have become poorer. Material consumption has also increased as a result, so there are also more resources being used to in the demands to produce/purchase these goods than in meeting the needs of the majority poor. This has further led to "blaming the victims" for their poverty, by some (not all, of course) of the wealthier classes of their own nation.

As the effects of globalization are creating further disparities and inequalities, around the world we are seeing an increase in violence and human rights abuses as disputes about territories, food and water are spilling into wars and internal conflicts. People are fighting for basic needs. The following are some of the areas of current and future tension.

— the various conflicts in Africa. It is also feared that conflicts involving water will increase.

— the Middle East where national interests in the vast oil fields have led to wars and influence from states like USA and UK.

— the 1998 riots in Indonesia fueled by the current global financial crisis.

— the Nile area, where Egypt rely on downstream water largely controlled by Ethiopia.

— Iraq, Syria and Turkey where there is tension surrounding the water flow of the Euphrates and Tigris.

— Israel and Jordan, where Israel cut water supplies to Jordan due to sever drought

— Israel and Palestine also are fighting over water resources as well.

— The Chiapas region in Mexico

— Water scarcity in the Gaza region has contributed to the tensions in the Middle East.

— Environmental scarcity and social tensions in Pakistan have led to a worsening situation.

- — Tensions in the Narmada region in India between indigenous people and the government.
- — Environmental scarcity in Rwanda contributed to the ethnic conflicts in 1994.
- — Degradation on the environment and an increase in population is fueling tension in South Africa.
- — In Equador, it is predicted that extreme violence is going to be seen at indigenous protests against giant oil corporations.

World Watch Institute also point out that water will once again be at the center of new conflicts. They point out that things like IMF and World Bank-backed privatization policies, flawed big damn projects etc have caused further tensions, protest and violence. Many from those conflicts above, and other conflicts not listed, also see the underlying cause of overly corporate-led globalization as a root cause as well, as they and the foreign policies of the wealthier nations have allowed economic and resource-controlling policies to be instituted in their favor. Also, while famine is often said to result from effects that are said to be caused by over-population, it is often overlooked on how the impact of politics and economics have a far more significant impact on famines than do "over" population and that those impacted would have a distinct class distinction.

Global Warming

Global warming and climate change refer to an increase in average global temperatures. Natural events and human activities are believed to be contributing to an increase in average global temperatures. This is caused primarily by increases in "greenhouse" gases such as Carbon Dioxide (CO_2).

Greenhouse Effect

The term greenhouse is used in conjunction with the phenomenon known as the greenhouse effect.

— Energy from the sun drives the earth's weather and climate, and heats the earth's surface;

— In turn, the earth radiates energy back into space;

— Some atmospheric gases (water vapor, carbon dioxide, and other gases) trap some of the outgoing energy, retaining heat somewhat like the glass panels of a greenhouse;

— These gases are therefore known as greenhouse gases;

— The greenhouse effect is the rise in temperature on Earth as certain gases in the atmosphere trap energy.

Six main gases considered to be contributing to global climate change are carbon dioxide (CO_2), methane (CH_4) and nitrous oxide (N_2O), plus three fluorinated industrial gases: hydrofluorocarbons (HFCs), perfluorocarbons (PFCs) and sulphur hexafluoride (SF_6). Water vapor is also considered a greenhouse gas. Many of these greenhouse gases are actually life-enabling, for without them, heat would escape back into space and the Earth's average temperature would be a lot colder. However, if the greenhouse effect becomes stronger, then more heat gets trapped than needed, and the Earth might become less habitable for humans, plants and animals.

Impacts of Global Warming

For decades, greenhouse gases, such as carbon dioxide have been increasing in the atmosphere. But why does that matter? Won't warmer weather be nicer for everyone? Increased greenhouse gases and the greenhouse effect is feared to contribute to an overall warming of the Earth's climate, leading to a global warming. Consider also the following:

— 1998, 2002 and 2003 have been the three warmest years on record;

— According to the World Meteorological Organization (WMO), the 1990s was the warmest decade;

— The 1900s was the warmest century during the last 1,000 years.

However, it is the rapid pace at which the temperature will rise that will result in many negative impacts to humans and the environment and this why there is such a world-wide concern. Most scientists believe that the warming of the climate will lead to more extreme weather patterns such as:

— More hurricanes and drought;

— Longer spells of dry heat or intense rain (depending on where you are in the world);

— Scientists have pointed out that Northern Europe could be severely affected with colder weather if climate change continues, as the arctic begins to melt and send fresher waters further south. It would effectively cut off the Gulf Stream that brings warmth from the Gulf of Mexico, keeping countries such as Britain warmer than expected;

— In South Asia, the Himalayan glaciers could retreat causing water scarcity in the long run.

While many environmental groups have been warning about extreme weather conditions for a few years, the World Meteorological Organization announced in July 2003 that "Recent scientific assessments indicate that, as the global temperatures continue to warm due to climate change, the number and intensity of extreme events might increase." The WMO also notes that "New record extreme events occur every year somewhere in the globe, but in recent years the number of such extremes have been increasing." The U.K's Independent newspaper described the WMO's announcement as "unprecedented" and "astonishing" because it came from a respected United Nations organization not an environmental group!

With global warming on the increase and species' habitats on the decrease, the chances for various ecosystems to adapt naturally are diminishing. Many studies have pointed out that the rates of extinction of animal and plant species, and the temperature changes around the world since the industrial revolution, have been significantly different to normal expectations. An analysis of population trends, climate change, increasing pollution and emerging diseases found that 40 percent of deaths in the world could be attributed to environmental factors.

Global warming will lead to the following situations, amongst others:

— Rapid global heating according to a US National Academy of Science warning;

— Dramatic increase in greenhouse gas emissions;

— Ozone loss aggravated by global warming;

— Ozone loss likely to aggravate global warming;

— Warming of the oceans leads to increased green house gasses;

— Permafrost thawing will aggravate global warming;

— Oceanic changes observed that may aggravate the situation;

— A vicious circle whereby each problem will exacerbate other problems which will feedback into each other;

— Massive exctinction of species will aggravate the environmental crisis;

— Sudden collapse of biological and ecological systems may occur, but will have a very slow recovery;

— While effective measures can decrease global warming and other problems the World community has repeatedly failed to establish cooperation.

Rising Sea Levels

Water expands when heated, and sea levels are expected to rise due to climate change. Rising sea levels will also result as the polar caps begin to melt. Rising sea levels is already affecting many small islands. Rising sea levels will impact many coastlines, and a large mass of humanity lives near the coasts or by major rivers.

An increase in pests and disease is also feared. Climate change is disrupting natural ecosystems in a way that is making life better for infectious diseases.

Leading climate scientists now agree that human pollution, mainly from fossil fuels, has added substantially to global warming in the past 50 years. A 1000-page document from the UN's Intergovernmental Panel on Climate Change (IPCC) reveals far bleaker predictions that the Earth will get far hotter than predicted. As the World Resources Institute highlights there is a huge contrast between developed/industrialized nations and poorer developing countries in greenhouse emissions, as well as the reasons for those emissions. For example:

— In terms of historical emissions, industrialized countries account for roughly 80% of the carbon dioxide buildup in the atmosphere to date. Since 1950, the U.S. has emitted a cumulative total of roughly 50.7 billion tons of carbon, while China (4.6 times more populous) and India (3.5 times more populous) have emitted only 15.7 and 4.2 billion tons respectively (although their numbers will rise).

— Annually, more than 60 percent of global industrial carbon dioxide emissions originate in industrialized countries, where only about 20 percent of the world's population resides.

— Much of the growth in emissions in developing countries results from the provision of basic human needs for growing populations, while emissions in industrialized countries contribute to growth in a standard of living that is already far above that of the average person worldwide. This is exemplified by the large contrasts in per capita carbons emissions between industrialized and developing countries. Per capita emissions of carbon in the U.S. are over 20 times higher than India, 12 times higher than Brazil and seven times higher than China.

The United States is the world's largest emitter of greenhouse gases. It:

— Accounts for roughly four percent of the world's population;

— Accounts for approximately 23% of global emissions and 42% of industrialized country emissions;

The previously 15 member-nations European Union (E.U.), if considered as a whole (for it is more comparable to the U.S.):

— Accounts for roughly 3 percent of the world's population;

— Accounts for around 10% of global emissions and 24% of industrialized countries' man-made emissions of the six main gases;

— Recent years have seen a reduction in emissions from those initial 15-member states. However,

a) It is not near the level required;

b) For the second consecutive year, in 2003, emissions from EU countries have actually increased slightly (though still remaining slightly lower than 1990 levels).

The Kyoto Protocol was the climate change treaty negotiated in 1997, setting targets for emissions of greenhouse gases. In order to be binding under international law, the treaty would need ratification from the countries responsible for around 55% of the global greenhouse gas emissions of 1990. The U.S. being the world's largest emitter of greenhouse gasese, pulled out in 2001, leaving treaty ratification depdenent on Russia, responsible for 17% of world emissions. Russia has to cut emission levels from the Soviet days, and their emissions in the past decade has been far less, so it should not pose as much of a problem to reduce such emissions.

Skepticism

For a very long time, something of contention and debate in the U.S. had been whether or not a lot of climate change has in fact been induced by human activities, while many scientists around the world, Europe especially, have been more convinced that this is the case. In May 2002, the Bush Administration in the U.S. did admit a link between human activities and climate change. However, at the same time the administration has continued its controversial stance of maintaining that it will not participate in the international treaty to limit global warming, the Kyoto Protocol, due to economic priorities and concerns.

Throughout the 1990s, especially in the United States, but in other countries as well, those who would try and raise the importance of this issue, and suggest that we are perhaps over-consuming, or unsustainably using our resources etc, were faced with a lot of criticism and ridicule. The previous link is to an article by George

Monbiot, writing in 1999. In 2004, he notes a similar issue, whereby media attempts at balance has led to "false balancing" whereby disproportionate time is given to more fringe scientists or those with less credibility or with additional agendas, without noting so, and thus gives the impression that there is more debate in the scientific community about whether or not climate change is an issue to be concerned about or not:

> Picture a situation in which most of the media, despite the overwhelming weight of medical opinion, refused to accept that there was a connection between smoking and lung cancer. Imagine that every time new evidence emerged, they asked someone with no medical qualifications to write a piece dismissing the evidence and claiming that there was no consensus on the issue. Imagine that the BBC, in the interests of "debate", wheeled out one of the tiny number of scientists who says that smoking and cancer aren't linked, or that giving up isn't worth the trouble, every time the issue of cancer was raised.
>
> Imagine that, as a result, next to nothing was done about the problem, to the delight of the tobacco industry and the detriment of millions of smokers. We would surely describe the newspapers and the BBC as grossly irresponsible. Now stop imagining it, and take a look at what's happening. The issue is not smoking, but climate change. The scientific consensus is just as robust, the misreporting just as widespread, the consequences even graver.

Despite the ridicule and attempts to discredit global warming concerns, there have been enough studies to conclude that global climate change is here. For example,:

— According to a study, by Patrick Mazza and Rhys Roth, there is enough evidence to conclude that Global Warming is here and human-induced. Their study is titled Global Warming is Here; The Scientific Evidence;

— A scientific study by the Imperial College of Science, Technology and Medicine, London, verified the warming of Earth's climate from space by comparison of satellite data over a 27 year time span.

Pollution from various industries, the burning of fossil fuels, methane from farm animals, forest destruction, rotting/dead vegetation etc have led to an increased number of greenhouse gases in the atmosphere. And, as international trade in its current form continues to expand with little regard for the environment, the transportation alone, of goods is thought to considerably contribute to global warming via emissions from planes, ships and other transportation vehicles.

Even sulphur emitted from ships are thought to contribute a fair bit to climate change. In fact, sulphur based gas, originating from industry, discovered in 2000 is thought to be the most potent greenhouse gas measured to date. It is called trifluoromethyl sulphur pentafluoride (SF_5CF_3). With Earth's resources gradually being depleted, sustainability and alternative technologies become even more important. While some major companies are even trying to produce more efficient products or use energy more efficiently, other large corporations are actually pushing back environmental programs in order to increase profits or to survive in a tough business world. The efforts of others to help protect the environment, and ultimately ourselves, are seriously undermined, as a result.

Global Dimming

Fossil fuel use, as well as producing greenhouse gases, creates other by-products. These by-products are also pollutants, such as sulphur dioxide, soot, and ash. These pollutants however, also change the properties of clouds. Clouds are formed when water droplets are seeded by air-borne particles, such as pollen. Polluted air results in clouds with larger number of droplets than unpolluted clouds. This then makes those clouds more reflexsive. More of the sun's heat and energy is therefore reflected back into space. This reduction of heat reaching the earth is known as Global Dimming.

Global warming results from the greenhouse effect caused by, amongst other things, excessive amounts of greenhouse gases in the earth's atmosphere from fossil fuel burning. It would seem then, that the other by-products which cause global dimming may be an ironic saviour. The pollutants that lead to global dimming also lead to various human and environmental problems, such as smog, respiratory problems, and acid rain.

The impacts of global dimming itself, however, can be devastating. The death toll that global dimming may have already caused is thought to be massive. Climatologists studying this phenomenon believe that the reflection of heat have made waters in the northern hemisphere cooler. As a result, less rain has formed in key areas and crucial rainfall has failed to arrive over the Sahel in Northern Africa.

In the 1970s and 1980s, massive famines were caused by failed rains which climatologists had never quite understood why they had failed. The answers that global dimming models seemed to provide, the documentary noted, has led to a chilling conclusion: "what came out of our exhaust pipes and power stations [from Europe and North America] contributed to the deaths of a million people in Africa, and afflicted 50 million more" with hunger and starvation.

Scientists said that the impact of global dimming might not be in the millions, but billions. The Asian monsoons bring rainfall to half the world's population. If this air pollution and global dimming has a detrimental impact on the Asian monsoons some 3 billion people could be affected.

Contrails were seen as another significant cause of heat reflection. During the aftermath of the September 11, 2001 terrorist attacks in the United States, all commercial flights were grounded for the next three days. This allowed climate scientists to look at the effect on the climate when there were no contrails and no heat reflection. What scientists found was that the temperature rose by some 1 degree centigrade in that period of 3 days.

The impacts of global dimming have led to fears that global dimming has been hiding the true power of global warming. Currently, most climate change models predict a 5 degrees increase in temperature over the next century, which is already considered extremely grave. However, global dimming has led to an underestimation of the power of global warming.

Global dimming can be dealt with by cleaning up emissions. However, if global dimming problems are only addressed, then the effects of global warming will increase even more. This may be what happened to Europe in 2003. In Europe, various measures have been taken in recent years to clean up the emissions to reduce pollutants that create smog and other problems, but without reducing the greenhouse gas emissions in parallel. This seems to have had a few effects:

— This may have already lessened the severity of droughts and failed rains in the Sahel.

— However, it seems that it may have caused, or contributed to, the European heat wave in 2003 that killed thousands in France, saw forest fires in Portugal, and caused many other problems throughout the continent.

The impacts of addressing global dimming only would increase global warming more rapidly. Irreversible damage would be only about 30 years away. Global level impacts would include:

— The melting of ice in Greenland, which would lead to more rising sea levels. This in turn would impact many of our major world cities

— Drying tropical rain forests would increase the risk of burning. This would release even more carbon dioxide into the atmosphere, further increasing global warming effects.

These and other effects could combine to lead to an increase of 10 degrees centigrade in temperature over the next 100 years, not the standard 5 degrees which most models currently predict. This would be a more rapid warming than any other time in history. With such an increase,

— Vegetation will die off even more quickly

— Soil erosion will increase and food production will fail

— A Sahara type of climate could be possible in places such as England, while other parts of the world would fare even worse.

— Such an increase in temperature would also release one of the biggest stores of greenhouse gases on earth, methane hydrate, currently contained at the bottom of the earth's oceans and known to destabilize with warming. This gas is eight times stronger than carbon dioxide in its greenhouse effect. As the documentary also added, due to the sheer amounts that would be released, by this time, whatever we would try to curb emissions, it would be too late.

Causes of Global Warming

If we were to use global dimming pollutants to stave off the effects of global warming, we would still face many problems, such as:

— Human health problems from the soot/smog
— Environmental problems such as acid rain
— Ecological problems such as changes in rainfall patterns which can kill millions, if not billions.

Climatologists are stressing that the roots of both global dimming causing pollutants and global warming causing greenhouse gases have to be dealt with together and soon.

3

Environmental Pollution in India

Water Pollution

When toxic substances enter lakes, streams, rivers, oceans, and other water bodies, they get dissolved or lie suspended in water or get deposited on the bed. This results in the pollution of water whereby the quality of the water deteriorates, affecting aquatic ecosystems. Pollutants can also seep down and affect the groundwater deposits. Water pollution has many sources.

The most polluting of them are the city sewage and industrial waste discharged into the rivers. The facilities to treat waste water are not adequate in any city in India. Presently, only about 10% of the waste water generated is treated; the rest is discharged as it is into our water bodies. Due to this, pollutants enter groundwater, rivers, and other water bodies.

Such water, which ultimately ends up in our households, is often highly contaminated and carries disease-causing microbes. Agricultural run-off, or the water from the fields that drains into rivers, is another major water pollutant as it contains fertilizers and pesticides. Domestic sewage refers to waste water that is discarded from households. Also referred to as sanitary sewage, such water contains a wide variety of dissolved and suspended impurities.

The amount of organic material that can rot in the sewage is measured by the biochemical oxygen demand. BOD is the amount of oxygen required by micro-organisms to decompose the organic substances in sewage. Therefore, the more organic material there is in the sewage, the higher the BOD. It is among the most important parameters for the design and operation of sewage treatment plants. BOD levels of industrial sewage may be many times that of domestic sewage. Dissolved oxygen is an important factor that determines the quality of water in lakes and rivers. The higher the concentration of dissolved oxygen, the better the water quality.

When sewage enters a lake or stream, micro-organisms begin to decompose the organic materials. Oxygen is consumed as micro-organisms use it in their metabolism. This can quickly deplete the available oxygen in the water. When the dissolved oxygen levels drop too low, many aquatic species perish. In fact, if the oxygen level drops to zero, the water will become septic. When organic compounds decompose without oxygen, it gives rise to the undesirable odours usually associated with septic or putrid conditions.It amounts to a very small fraction of the sewage by weight. But it is large by volume and contains impurities such as organic materials and plant nutrients that tend to rot.

The main organic materials are food and vegetable waste, plant nutrient come from chemical soaps, washing powders, etc. Domestic sewage is also very likely to contain disease-causing microbes. Thus, disposal of domestic waste water is a significant technical problem. Sewage generated from the urban areas in India has multiplied manifold since 1947. Today, many people dump their garbage into streams, lakes, rivers, and seas, thus making water bodies the final resting place of cans, bottles, plastics, and other household products.

The various substances that we use for keeping our houses clean add to water pollution as they contain harmful chemicals. In the past, people mostly used soaps made from animal and vegetable fat for all types of washing. But most of today's cleaning products are synthetic detergents and come from the petrochemical industry. Most detergents and washing powders contain phosphates, which are used to soften the water among other things. These and other chemicals contained in washing powders affect the health of all forms of life in the water.

Eutrophication

When fresh water is artificially supplemented with nutrients, it results in an abnormal increase in the growth of water plants. This is known as eutrophication. The discharge of waste from industries, agriculture, and urban communities into water bodies generally stretches the biological capacities of aquatic systems. Chemical run-off from fields also adds nutrients to water. Excess nutrients cause the water body to become choked with organic substances and organisms.

When organic matter exceeds the capacity of the micro-organisms in water that break down and recycle the organic matter, it encourages rapid growth, or blooms, of algae. When they die, the remains of the algae add to the organic wastes already in the water; eventually, the water becomes deficient in oxygen. Anaerobic organisms (those that do not require oxygen to live) then attack the organic wastes, releasing gases such as methane and hydrogen sulphide, which are harmful to the oxygen-requiring (aerobic) forms of life.

The result is a foul-smelling, waste-filled body of water. This has already occurred in such places as Lake Erie and the Baltic Sea, and is a growing problem in freshwater lakes all over India. Eutrophication can produce problems such as bad tastes and odours as well as green scum algae. Also the growth of rooted plants increases, which decreases the amount of oxygen in the deepest waters of the lake. It also leads to the death of all forms of life in the water bodies.The use of land for agriculture and the practices followed in cultivation greatly affect the quality of groundwater.

Intensive cultivation of crops causes chemicals from fertilizers (e.g. nitrate) and pesticides to seep into the groundwater, a process commonly known as leaching. Routine applications of fertilizers and pesticides for agriculture and indiscriminate disposal of industrial and domestic wastes are increasingly being recognized as significant sources of water pollution. The high nitrate content in groundwater is mainly from irrigation run-off from agricultural fields where chemical fertilizers have been used indiscriminately.

Industrial Effluents

Waste water from manufacturing or chemical processes in industries contributes to water pollution. Industrial waste water usually contains specific and readily identifiable chemical compounds. During the last fifty years, the number of industries in India has grown rapidly. But water pollution is concentrated within a few subsectors, mainly in the form of toxic wastes and organic pollutants. Out of this a large portion can be traced to the processing of industrial chemicals and to the food products industry.

In fact, a number of large- and medium-sized industries in the region covered by the Ganga Action Plan do not have adequate effluent treatment facilities. Most of these defaulting industries are sugar mills, distilleries, leather processing industries, and thermal power stations. Most major industries have treatment facilities for industrial effluents. But this is not the case with small-scale industries, which cannot afford enormous investments in pollution control equipment as their profit margin is very slender.

Effects of Water Pollution

The effects of water pollution are not only devastating to people but also to animals, fish, and birds. Polluted water is unsuitable for drinking, recreation, agriculture, and industry. It diminishes the aesthetic quality of lakes and rivers. More seriously, contaminated water destroys aquatic life and reduces its reproductive ability. Eventually, it is a hazard to human health. Nobody can escape the effects of water pollution. The individual and the community can help minimize water pollution. By simple housekeeping and management practices the amount of waste generated can be minimized.

Ground Water Use

It is estimated that 80% of domestic needs in rural areas and 50% in urban areas is met by ground water.India's total replenishable groundwater have been estimated at 431.8 kmby the Central Statistical Organisation. The average level of groundwater development in India is 32%, although some states have exploited their resources to a much greater extent (94% in Punjab, 84% in Haryana, 60% in Tamil Nadu, 64% in Lakshadweep, 51% in Rajasthan). 85% of ground water extracted is used for irrigation purposes and 15% for Industrial and domestic. purposes.Reciprocally, as much as 70 to 80% of India's agricultural output may be groundwater dependent.

Fluoride

High fluoride concentration in ground water beyond the

permissible limit of 1.5 p.p.m is one of the most worrying toxicological problems India. This contamination has geological origins, and dangerous level of Fluoride presence in ground water have been recorded in 17 different states, covering almost the hole territory. In affected areas, the range of fluoride contamination varies between 1.5 and 29 p.p.m. Prolonged ingestion of high quantities of fluoride can lead to dental or skeletal fluorosis. According to a survey carried out by the Rajiv Gandhi National Drinking Water Mission (RGNDWM) in 1993, around 25 million people were affected at this time. The population at risk is estimated at 66 million

The fluoride is present in many different geological formations, but the gneiss of the Indian underground have been identified as being responsible for the problematic fluoride concentration in ground water. The main methods identified for Fluoride removal are

— Co-precipitaion
— Adsorption by activated carbon and activated alumina
— Exchange method
— Reverse Osmosis

Arsenic

Arsenic contamination of groundwater has an unusually high and dramatic occurrence in the gangetic delta including West Bengal and Bangladesh. The first indication of arsenic-related health problems was reported in West Bengal in 1983. It took ten more years for the situation to be recognised in Bangladesh. Since then, several programs of investigation have been carried out and the arsenic crisis in the ganga delta is now the focus of many efforts worldwide to release the population living in this area from this serious threat. Arsenic contamination through drinking water may be responsible for cancer of skin, lungs, urinary bladder, and kidney, as well as other skin affectations. The population at risk is estimated at 5.3 million, with 200000 people actually affected.

Arsenic is widespread in the earth's crust. The concentration of arsenic, however, is particularly high in the thick succession of fluviatile sediments pertaining to the quaternary age that form the majority of the exploited aquifers in the region. Although everyone agrees about the geologic origin of the arsenic that can be found in ground water in West Bengal and Bangladesh, there is no consensus about the reason why the arsenic gets diluted at such a high rate.

Some scientist assume that the high arsenic concentration in groundwater can be explained by the exposition of arseno-ferous complex to atmospheric oxygen introduced in the aquifer in response to lowering of groundwater level. An alternative explanation puts forward the strongly reducing condition of the aquifer, that can be explained by the burial of organic sediment during its geological formation. This two - mostly geogenic - explanations are challenged by another one based on the action of phosphates from chemical fertilizers that could displace arsenic from the sediment.

WHO standards concerning Arsenic concentration in drinking water established in 1993 an allowable concentration of 0.01mg/L, bringing down the 1963 limit of 0.05mg/L. However, the permissible limit in India and Bangladesh, is still 0.05 mg/L, and most of the test done in those areas use this reference. Nine of sixteen districts of West Bengal have been reported to have ground water arsenic concentrations above 0.05mg/L. According to the Public Health Engineering Department (PHED) in Calcutta, the population of those nine districts is estimated at 39 million.

There are around 22000 public tubewells and more than 400000 private tubewells in the area. In the tenth five-year plan document prepared by the Indian Planning Commission, it was reported that arsenic contaminated habitations had been identified in eight districts, with 200,000 people actually affected, and an exposed population around 5.3 million. The reported number of tubewells in the affected area was 22000 public tubewells, and 130000 private tubewells. These figures, compared to the one provided by the PHED, show the lack of information about the private tubewells, which is a major obstacle in the efforts to assess the real extent of the crisis.

The problem of arsenic in drinking water is maybe the more urgent to be solved, but finding solutions for providing safe drinking water will not prevent totally As ingestion as long as As enters the surface environment through extracted ground water. Animals drinking contaminated water may accumulate As in their tissues, as well as crops grown in a field irrigated with As contaminated groundwater. Irrigation poses an even more serious threat to the environment, through a potential long term contamination of soils, and maybe a subsequent contamination of surface water flows.

Until recently, no study was conducted to explore those alternative pathways. The Indian Council of Agricultural Research as started a study in 1998 about the impact of As contamination in terms of agricultural management.Although the surface water was not found to be heavily contaminated, the study showed that As tends to accumulate in most of the crop cultivated on the soils irrigated with contaminated water. Moreover, the As present in the crops is mostly in its most toxical form of arsenate. Providing an arsenic-free water supply, can be achieved by :

— Using surface water.

— Rainwater harvesting

Tapping into deeper aquifers (100-150 meters), or shallow aquifers (hand-dug wells less deep than 20 meters) which have been found to be arsenic free. Providing an arsenic removal technology which can be used, either at the level of the household, either at a more central level. There is now eight main classes of arsenic remediation technologies based on the following processes:

1. Oxidation,
2. Coagulation/Coprecipitation,
3. Sedimentation,
4. Filtration,
5. Adsorption,
6. Ion Exchange,
7. Membrane/Reverse Osmosis,
8. Biological.

The Water and Sanitation Program (WSP) has conducted a field study about the technologies being tested in West Bengal and Bangladesh for Arsenic Mitigation. Seven promising technologies were identified.

Iron

High concentration of Iron in drinking water is the most widespread quality problem in India. Nevertheless, the health impact of this contamination is smaller than the one caused by arsenic of fluoride contamination. Using Iron contaminated water, that is over the permissible limit of 1 ppm can cause constipation accompanied by other physiological disorders. The RGNDWM sub mission dedicated to the removal of Excess Iron had approved 1615 and 9355 plants had already been commissioned in 1998.

Brackishness

Brackishness may be an initial characteristic of some aquifers, but it can be caused or worsened by infiltration due to overexploitation of the aquifer. This is often the case in coastal areas where aquifers get contaminated with infiltrated sea water. Excess brackishness in drinking water has laxative effects. The permissible limit is set at 1500 ppm of total dissolved solids (TDS). 194 desalination plants have been approved by the RGDWM and 150 plants had been commissioned up to 1998.

Air Pollution

In many people's minds air pollution is associated with the contamination of urban air from automobile exhausts and industrial effluents. However, in developing countries, the problem of indoor air pollution far outweighs the ambient air pollution. There are four principal sources of pollutants of indoor air:

(i) combustion,

(ii) building material,

(iii) the ground under the building, and

(iv) bioaerosols.

In developed countries the most important indoor air pollutants are radon, asbestos, volatile organic compounds, pesticides, heavy metals, animal dander, mites, moulds and environmental tobacco smoke. However, in developing countries the most important indoor air pollutants are the combustion products of unprocessed solid biomass fuels used by the poor urban and rural folk for cooking and heating. Approximately half the world's population and up to 90% of rural households in developing countries still rely on unprocessed biomass fuels such as wood, dung and crop residues.

A recent report of the World Health Organization (WHO) asserts the rule of 1000 which states that a pollutant released indoors is one thousand times more likely to reach people's lung than a pollutant released outdoors. It has been estimated that about half a million women and children die each year from indoor air pollution in India. Compared to other countries, India has among the largest burden of disease due to the use of dirty household fuels and 28% of all deaths due to indoor air pollution in developing countries occur in India.

The type of fuels used by a household is determined maily by its economic status. In the energy ladder, biomass fuels namely animal dung, crop residues and wood, which are the dirtiest fuels, lie at the bottom and are used mostly by very poor people. Electricity, which is the most expensive, lies at the top of ladder and it is also the cleanest fuel. The 1991 National Census for the first time inquired about the fuel used for coking. It revealed that about 90% of the rural population relied upon the biomass fuels like animal dung, crop residues and wood. A small portion used coal. Nation-wide about 78% of the population relied upon the biomass fuels and 3% on coal.

Air Pollutants Released from Biomass Combustion

It has been estimated that more than half world's households cook their food on the unprocessed solid fuels that typically release at least 50 times more noxious pollutants than gas. The stoves or chullah used for cooking are not energy efficient. The fuels re not burned completely. The incomplete combustion of biomass releases complex mixture of organic compounds, which include suspended particulate matter, carbon monoxide, poly organic material (POM), poly aromatic hydrocarbons (PAH), formaldehyde, etc. The biomass may also contain intrinsic contaminants such as sulphur, trace metals, etc.

In recent years a large number of studies of health impact of suspended particulate air pollution have been undertaken in developing countries. These studies show remarkable consistency in the relationship observed between changes in daily ambient suspended particulate levels and changes in mortality. Smithestimated the health risk from exposure to particulate air pollution by applying the mean risk per unit ambient concentrations based on the results of some urban epidemiological studies.

The range of risk was found to be 1.2 - 4.4% increased mortality per 10 mg/mincremental increase in concentration of respirable suspended particles (PM). For the calculations of estimates, it was assumed that the health risk has linear relationship to exposure, the risk factors determined for urban centers of developed nations were used as standards; where the PMdata were not available, 50% of suspended particulate matter (SPM) levels were considered as equivalent. The above assumptions may add to inaccuracy already inherent in such estimates.

Carbon Monoxide

Incomplete combustion of fuels produces carbon monoxide (CO). The CO and particle emission pose a serious problem when biomass fuels are used. Smithhas estimated that about 38, 17, 5 and 2 g/meal carbon monoxide is released during the household cooking, using dung, crop residues, wood and kerosene respectively. During the use of liquid petroleum gas (LPG) a negligible amount of CO is released.

A studyby the National Institute of Occupational Health (NIOH), Ahmedabad reported indoor air CO levels of 144, 156, 94, 108 and 14 mg/mair during cooking by dung, wood, coal, kerosene and LPG respectively. The short-term health effects of CO exposure are dizziness, headache, nausea, feeling of weakness, etc. The association between long-term exposure to carbon monoxide from cigarette smoke and heart disease and foetal development has been described by several authors.

Poly Organic Material and Poly Aromatic Hydrocarbons

Poly organic material is a loose term used to depict a group of chemicals having two or more rings. Of several chemicals included in this group, the PAHs have attracted interest for their possible carcinogenic effects. In addition to PAH, azo and arino compounds have also been found to be potentially carcinogenic. Most other categories of POM are of less environmental interest or are not found in large amounts in organic combustion products.

Polycyclic aromatic hydrocarbons constitute a large class of compounds released during the incomplete combustion or pyrolysis of organic matter. They are often called polynuclear aromatic (PNA) because they contain three or more aromatic rings that share carbon atoms. Benzo(a)pyrene (BaP) is one of the most important carcinogen of the group. Often it is measured to indicate the presence or absence of PAHs although the relationship between BaP content and actual carcinogenicity may be weak.

Anthracene and phenanthracene are not carcinogens but methyl additions may render them carcinogenic. PAHs are activated by the hepatic microsomal enzyme system to carcinogenic forms that bind covalently to DNA. Study by NIOHshowed that the indoor levels of PAH (total) during use of dung, wood, coal, kerosene and LPG were 3.56, 2.01, 0.55, 0.23 and 0.13 µg/mof air respectively. These PAH were fluorene, pyrene, chrysene, benzo(a)anthracene, benzo(b)fluoranthene, benzo(k)fluoranthene, benzo(a)pyrene, dibenz(ah) anthracene, benzo(ghi)perylene and indeno(1,2,3-cd) pyrene. All these PAHs except the first three have been classified as possible carcinogens.

Formaldehyde

Patel and Raiyanimeasured levels of formaldehyde in indoor environment during cooking by different fuels. The formaldehyde mean levels were 670, 652, 109, 112 and 68 μg/mof air for cattle dung, wood, coal, kerosene and LPG respectively. The formaldehyde is well recognized to be an acute irritant and long-term exposure can cause a reduction in vital capacity and chronic bronchitis. The formaldehyde is well known to form crosslinks with biologic macro-molecules. Inhaled formaldehyde forms DNA and DNA-protein cross-links in the nasal respiratory mucosa.

The formaldehyde has been shown to be carcinogenic in a dose dependent fashion in rodents. The studies done in workers occupationally exposed to formaldehyde have consistently (11 of 13 studies reviewed) shown higher incidence of leukaemia. In an epidemiological study in U.K., significantly excess mortality from lung cancer was observed in workers exposed to high levels of formaldehyde.

Mutagenic Activities

Microbial tests are widely used as a screening tool for assessing mutagenic potential of chemical substances. The particulate matter in the smoke generated as a result of incomplete combustion of biomass fuels contains a number of organic compounds. To evaluate their carcinogenic potential, it is necessary to screen their mutagenicity through simple and rapid microbial assay as a first step. Ames assay is simple and sensitive enough to measure mutagenicity of air-borne particulates, so that many researchers have applied this assay to demonstrate the ambient carcinogenic and mutagenic compounds in the extractable organic matter from air-borne particulates.

Mutagenic response of complex mixtures of polycyclic organic matter from the combustion of biomass energy fuels was studied using tester strains TA 98 and TA 100 of Salmonella typhimurium which can detect the presence of frame-shift and base-pair mutagens. The results indicated that the organic residues of smoke particulates of wood and cattle dung fuels contained direct acting frameshift mutagens and cattle dung contained only direct acting base-pair mutagens while indirect acting frame-shift and base-pair mutagens were found to present in smoke particulates of both the energy fuels.

Diseases Caused by Air Pollution

Respiratory illness, cancer, tuberculosis, perinatal outcomes including low birth weight, and eye diseases are the morbidities associated with indoor air pollution.

Respiratory Illness: The effect of air pollutants in general would depend on the composition of the air that is inhaled which will depend on the type of fuel used and the conditions of combustion, ventilation and duration for which the inhalation occur. The most commonly reported and obvious health effect of indoor air pollutants is the increase in the incidence of respiratory morbidity. Studies by the NIOHon the prevalence of respiratory symptoms in women using traditional fuels (biomass) (n=175) and LPG (n=99), matched for economic status and age, indicated that the relative risk (with 95% C.I.) for cough, and shortness of breath (dyspnoea) was 3.2 (1.6-6.7), and 4.6 (1.2-18.2) respectively.

Lung Cancer: The link between lung cancer in Chinese women and cooking on an open coal stove has been well established. Smoking is a major risk factor for lung cancer, however, about two-thirds of the lung cancers were reported in nonsmoking women in China, Indiaand Mexico. The presence of previous lung disease, for example tuberculosis which is common in Indian women, is a risk factor for development of lung cancer in non-smokers. The smoke from biomass fuels contain a large number of compounds such as poly aromatic hydrocarbons, formaldehyde, etc. known for their mutagenic and carcinogenic activities, but there is a general lack of epidemiological evidence connecting lung cancer with biomass fuel exposure.

The factors associated with rural environment may have a modulating effect on the occurrence of lung cancer and therefore the low incidence of lung cancer in Indian women should not lead to a final conclusion of no link between biomass exposure and lung cancer. It may be concluded that at present there is limited evidence of indoor exposure from coal fires leading to lung cancer and there is no evidence for the biomass fuels. Further investigations are needed to reach definite conclusions.

Pulmonary Tuberculosis: Mishra et alrecently reported the association between use of biomass fuels and pulmonary tuberculosis on the basis of analysis of data collected on 260,000 Indian adults interviewed during the 1992-93 National Family Health Survey. Persons living in households burning biomass fuels were reported to have odd ratio of 2.58 (1.98-3.37) compared to the persons using cleaner fuel, with an adjustment for confounding factors such as separate kitchen, indoor overcrowding, age, gender, urban or rural residence and caste.

There is experimental evidence to show that the exposure to wood smoke may increases susceptibility of the lungs to infections. Exposure to smoke interferes with the mucociliary defences of the lungsand decreases several antibacterial properties of lung macrophages, such as adherence to glass, phagocytic rate and the number of bacteria phagocytosed.

Chronic exposure to tobacco smoke also decreases cellular immunity, antibody production and local bronchial immunity, and there is increased susceptibility to infection and cancer. Indeed, tobacco smoke has been associated with tuberculosis. Although the evidence in favour of tuberculosis associated with biomass fuel exposure is extremely weak, there is a theoretical possibility of such an association and considering the public health importance of the problem further experimental and epidemiological studies are necessary.

Cataract: During cooking particularly with biomass fuels, air has to be blown into the fire from time to time especially when the fuel is moist and the fire is smouldering. This causes considerable exposure of the eyes to the emanating smoke. In a hospital-based case-control study in Delhi the use of liquefied petroleum gas was associated with an adjusted odds ratio of 0.62 (0.4-0.98) for cortical, nuclear and mixed, but not posterior sub capsular cataracts in comparison with the use of cow dung and wood.

An analysis of over 170,000 people in Indiayielded an adjusted odds ratio for reported partial or complete blindness of 1.32 (1.16-1.50) in respect of persons mainly using biomass fuel compared with other fuels after adjusting for socio-economic, housing and geographical variables; there was a lack of information on smoking, nutritional state, and other factors that might have influenced the prevalence of cataract. It is believed that the toxins from biomass fuel smoke are absorbed systematically and accumulate in the lens resulting in its opacity. The growing evidence that environmental tobacco smoke causes cataracts is supportive.

Adverse Pregnancy Outcome: Low birth weight (LBW) is an important public health problem in developing nations attributed mainly to undernutrition in pregnant women. Low birth weight has serious consequences including increased possibility of death during infancy. Exposure to carbon monoxide from tobacco smoke during pregnancy has been associated with LBW.

4

Environment Management

The regulatory and institutional decision-making framework for environmental protection in India is embodied in nine major acts of the Indian Parliament. These are: the Water (Prevention and Control of Pollution) Act of 1974 which established the Central Pollution Control Board (CPCB) and State Pollution Control Boards (SPCBs); the Air (Prevention and Control of Pollution) Act of 1977 which added the monitoring of air emissions to the responsibilities of the various Boards; the Environment (Protection) Act of 1986; the Forest (Conservation) Act of 1980, amended in 1988; the Motor Vehicle Act of 1938, amended in 1988; the Public Liability Insurance Act of 1991; and Notifications on the Coastal Regulation Zone, 1991; and Environmental Impact Assessment of Development Projects, in 1994, the National Environment Appellate Authority Act, 1997.

Most of the above Acts and Notifications are aimed at strengthening the command-andcontrol regime. New initiatives, especially in the form of a mix of regulations and legislation, fiscal incentives for technology acquisition, voluntary agreements, educational programs and information campaigns are required. Although the government has introduced some of these measures, more is required because the regulatory structure of a central authority, the ministry of environment and forests (and other ministries) and the Central Pollution Control Board linked to state- level implementation agencies have proved to be largely unsuccessful in effectively managing the protection of the environment.

An assessment of India's environmental management system suggests that weaknesses are evident at every administrative level at the center, state and district. Further, despite aid under specific investment projects, more needs to be done in the area of industrial pollution, particularly air emissions, coastal zone management, urban land use including the citing of industries, and mitigating environmental degradation in the mining sector.

In 1993, the government, through the ministry of environment and forests issued the Environment Action Program - India (EAP), developed after consultations with other ministries, states and NGOs. The EAP identified seven environmental priorities for India:

(a) conservation and sustainable utilisation of biodiversity in selected ecosystems including forests, mangroves, wetland, coral reefs, mountain ecosystems;

(b) afforestation, wastelands development and conservation of soil and moisture and ensuring that water sources are not polluted;

(c) control of industrial and related pollution with an accent on the reduction and or management of wastes, particularly hazardous wastes;

(d) improving access to clean technologies;

(e) tackling urban environmental issues;

(f) an alternative energy plan; and,

(g) strengthening scientific understanding of environmental issues, as well as strengthening structures for training at different levels, orientation and creating environmental awareness.

A1996 study indicated that (a) despite a strong legal framework and various ministries at the center, departments and boards at the state level; and local agencies involved in environmental management, implementation remains weak. Institutional capacity building to strengthen monitoring, enforcement, and compliance with existing laws can have a high payoff; and (b) the government alone cannot be responsible for environmental management, stronger efforts have to be made to involve the large nongovernmental organisation community in programs ranging from biodiversity conservation to alternative energy programs. The study also recognised that this cooperation will require transparent policies and practices such as early involvement in the environmental impact assessment process, access to information, and actions to increase involvement through public participatory processes.

In the Approach Paper to the Tenth Five Year Plan for the period 2002-2007, the government has recognised that mere economic growth is not enough and that development must be defined in broader terms of human being enhancement. The Tenth Plan has thus outlined a set of monitorable targets to achieve its objectives:

— Reduction of poverty ratio by 5 percentage points by 2007 and by 15 percentage points by 2012;

— Providing gainful high-quality employment to the labour force over the tenth Plan period;

— All children in school by 2003; all children to complete 5 years of schooling by 2007;

— Reduction of gender gaps in literacy and wages rates by at least 50 per cent by 2007.

— Reduction in decadal rate of population growth between 2001 and 2011 to 16.2 per cent · Increase in literacy rate to 75 per cent with in the Plan period;

— Reduction of Infant Mortality Rate (IMR) to 45 per 1000 live births by 2007 and to 28 by 2012.

— Increase in forest and tree cover to 25 per cent by 2007 and 33 per cent by 2012. · All villages to have sustained access to portable drinking water within the Plan period.

— Cleaning of major polluted rivers by 2007 and other notified stretches by 2012. As indicated, these targets relate to good environmental performance and governance, highlighting how cross-cutting environment management issues are.

The basic objective of the Tenth Plan is to bring about all round enhancement of human well being by eradicating poverty through adopting well conceived development strategies in which environmental concerns are posited as the vital aspects. environmental management and economic development are mutually supportive aspects of the same agenda. A poor environment undermines development, while inadequate development results in a lack of resources for environmental protection. Environmental conservation must go hand in hand with economic development because any economic development which destroys the environment will create more poverty, unemployment and disease.

Environmentally destructive economic development will impoverish the poor even further and destroy their livelihood resource base. Environmental concern must be linked to people's lives and well being, impact as these do various aspects of human existence tha t include health and livelihood amongst others. Air pollution, soil degradation, deforestation, desertification, shrinking wetlands.

Inadequate public health and sanitation, indoor pollution in rural areas, growing water scarcity, falling ground water tables, the lack of minimum flow in the rivers and over extraction of water for irrigation purposes are some of the environmental problems that need to be addressed first before any poverty alleviation programme can meet with success. Based on the objectives of the Approach Paper To The Tenth Five Year plan 2002-2007, the ministry of environment and forests prepared a strategy for the period and for the year 2002-2003.

Ministry of Environment and Forests

The ministry of environment and forests is charged with the responsibility of planning, promoting, coordination and overseeing the implementation of various environmental and forestry programmes. Responsibilities include environmental management to promote health considerations, focus on poverty alleviation by enhancing access of the poor to natural resources for livelihood and heightening awareness regarding environmentally sound living process by focusing on nature-human synergy. The ministry has been designated as the nodal agency in the country for the United Nations Environment Programme (UNEP), International Centre for Integrated Mountain development (ICIMOD) and looks after the follow-up of the United Nations' Conference on Environment and Development (UNCED).

The objectives are supported by a set of legislative and regulatory measures aiming at preservation, conservation and protection of environment as indicated above. The activities of this ministry can be broadly divided into four sub-sectors namely environment, forestry and wildlife, National Afforestation and Ecodevelopment Board (NAEB) and the National River Conservation Directorate (NRCD).

The ministry has identified the following strategy to meet the objectives of the Tenth Plan:

(a) To promote 'sustainable development' by emphasising the intrinsic linkage between environment management and socio- economic development of the people, through increased access to natural capital which is an essential component of reducing poverty. The emphasis will be on creating an enabling environment in terms of policies, institutions which support the over-arching objective of poverty reduction

(b) To mainstream environment issues in the developmental strategy, a multipronged approach will be followed which will include the incorporation of environment in the curriculum, sharpening the mechanism for the environmental auditing of the development activities, creating heightened awareness, broadbasing community involvement and to forging a coordinated effort with corporate and public sector in environmental management

(c) To input environmental concerns in the developmental scenarios by treating environment as a cross-sectoral issue

(d) To take concrete steps for increasing forest/tree cover from the existing level of 19.37 per cent to 25 per cent by 2007 and 33 per cent by 2012. While striving for this goal the thrust will be given to:

(1) increasing productivity;

(2) efficient management of resources; and,

(3) focus on the local needs and location specific strategies in terms of species selection and management of resources.

(e) Cleaning of major polluted rivers by 2007 and other notified stretches by 2012. In order to have efficient, smooth and time bound attainment of the goal, a holistic and integrated approach of addressing not only the intercepting sewers and sewage treatment plants which currently constituted the core of the Plan, but also other components as internal sewage, solid waste management and toilets with the participation of states, public and local bodies. The focus will on be treating the town as a unit by way of convergence of various schemes.

(f) Major changes in the institutional framework with a view to make it responsive and accountable to perceived needs of people

(g) Involvment of youth in environmental education, through a National Green Corps scheme to be implemented in all educational institutions covering class 8th to 12th class.

The thrust areas under the environmental sector for the Tenth Plan will be environmental impact assessment and abatement of pollution; ecological survey; conservation and eco-regeneration; environmental education, awareness and people's participation; environmental research and training; and, national river conservation. Under forestry, the thrust on sustainable management of forestry resources with people's participation initiated during the Eighth Plan and carried through the Ninth Plan will be further strengthen during the Tenth Plan period.

The programmes/schemes will focus on farm forestry, measures to sustain joint forest management, take care of the rights and interests of the tribal population living in an around the forests, review of the subsidies in the sale of non timber forest produce (NTFP) and promote bamboo cultivation and coastal shelter belt plantations for prevention of natural calamities. The wildlife programmes/schemes will focus on strengthening and enhancing the protective area networking through suitable policy framework and management plans; settlement of right and relocation of villages; promotion of ecodevelopment by giving priority to national parks and sanctuaries having higher biodiversity and high threats; timely payments of compensation against the depredation by wildlife; control of poaching; taxidermy and illegal trade in wildlife species; and, promotion of wildlife tourism.

The Tenth Plan thrusts for research and education segment will be on watershed management, desert development, medicinal plants, coastal areas, bamboos and agro forestry and biodiversity conservation. Added emphasis will be given to educational and training programmes.

The principal objective of environment sector for the Tenth Five Year Plan is to integrate the objectives of the policies stressed by the Planning Commission and to take cognizance of long term developmental perspectives related to industrialisation, power generation, transportation, agriculture, irrigation and other economic activities. Some of these goals have been identified as indicated below.

Environment

A. *Industrial pollution control*: It is proposed to ensure that large and medium polluting industrial units and stand alone polluting small scale industrial units (SSIs) install and operate pollution control devices and meet the norms. It will be ensured that small industries located in clusters and in industrial estates are covered under common effluent treatment plants (CETPs). Siting of pollution industries will be insisted to conform to the zoning plan which would be integrated with state industrial policy. Action plans will be effectively implemented to reduce pollution in all the critically polluted areas.

B. *Water pollution control*: Most of the rivers, streams, and large stretches of the coastal marine environment in

India are highly polluted with municipal waste, waste ge nerated from industry, chemical agents from fertilisers, pesticides from crop protection, and silt from degraded catchments. Untreated sewage and other non- industrial wastes account for four times as much pollution as industrial effluent. In the major cities, less than 50 per cent of the total wastewater generated is collected and less than one- fourth is treated. The consequence of such pollution is high levels of waterborne diseases which account for two-thirds of all illnesses in India with a significant loss in productivity.

Although there is a Coastal Regulation Zone Notification designed to protect the coastal marine ecosystems, implementation is weak and marine pollution continues largely unabated. Close monitoring of the discharge of industrial effluents into waterbodies will be enforced. A comprehensive plan is proposed to be prepared for setting up of sewage treatment facilities in 1,591 towns having populations more than 20,000). Overextraction of groundwater will be regulated. Areas of groundwater pollution will be identified and a plan will be implemented to contain this.

C. *Air pollution control*: Air quality in India's major cities indicate that ambient levels of nitrous oxides, sulphur dioxide, lead and suspended particulate matter are often higher than World Health Organization and Indian standards. Sulfur dioxide levels in nine of the country's major cities exceed national standards. Other harmful substances such as ozone are not monitored. Major contributors to air pollution problems in India include thermal power stations, industrial factories, vehicles, and the use of non-commercial fuels such as coal briquettes, animal dung and trash by large sections of the population. The negative environmental effects from increased fossil fuel use will likely increase as India seeks to meet the energy demands of its expanding economy. Coal will play a predominant role in this expansion. The major pollutants from coal exploitation include the principal green house gases (GHGs), particulate air emissions, coal mining runoff and wastes, and coal ash solid wastes which pose a large-scale disposal problem.

In view of the fact that cooperation of the state government is an essential factor in successfully implementing pollution control programmes, it is proposed to work with the state government for preparing location specific plans for reducing automobile pollution, and prevention of burning of biomass/garbage or any other material (except approved fuel) in the cities and towns to reduce secondary sources of air pollution. It is also proposed to set up a national network of air pollution monitoring stations for the parameters notified in respect of ambient air quality standards and for parameters for which standards have been prescribed for various sources. A database of pollutants releases into the air from different sources will be built up.

D. *Strengthening of central and state pollution controls board:* The board targets/activities proposed under the above are the training of officers, up-gradatio n of laboratories, capacity building in state pollution control boards / pollution control centres of the Northeast and in the new states of Uttranchal, Chattisgarh and Jharkhand. Economic instruments for prevention and control of pollution will be evolved. To implement the various economic instruments, specific proposals will be formulated keeping in view the relevant considerations which may require a separate set up.

E. *Environmental impact of human health* : Environmental studies will be initiated to assess the impact of various pollutants on human health. Measures will be taken to control/eliminate major source of pollutants which have an adverse effect on human health. Phasing out of highly polluting obsolete technologies is also envisaged during the Tenth Plan period.

F. *Hazardous substance management*: Comprehensive national chemical profiles are proposed to be prepared. Emergency response centers will be established in all industrial pockets with a large number of accident hazards units as also inventorisation of isolated storage of hazardous chemicals. Some of the other goals set for the Tenth Plan include effective implementation of the Public Liability Insurance Act; preparation of database on risk-prone areas; setting up of hard hazardous-based disposal facilities in all states where significant quantities of hazardous waste are being generated; establishment of sub-regional training centres for implementation of the Basel Convention, setting up of custom laboratories and capacity building of custom officers for checking contraband trade in hazardous chemicals/waste; implementation of environmentally sound practices in ship dismantling and proper management of biomedical wastes; municipal solid waste; plastic waste; and, chemical waste.

G. *Soil contamination and environmental degradation*: Contaminated sites are proposed to be identified and soil remediation plans will be prepared. Reclamation of degraded areas is another programme envisaged.

H. *Environmental impact assessmen (EIA)*: Capacity building will be undertaken for decentralised EIA of projects and monitoring of compliance of clearance condition. Detailed shoreline management plans and Coastal Zone Management plans are proposed to be prepared as also strengthening of state coastal zone management authorities. Plans for conservation of sensitive ecosystems will be prepared and implemented. Scoping of mega projects and strategic EIA for specific sectors will be undertaken.

I. *Ecological survey, conservation and ecogeneration*: Ongoing programmes are proposed to be strengthened and a status paper on achievement indicating the extent to which efforts have yielded the results will be brought out. Added efforts will be made to conserve biodiversity for ecological security including proper evaluation of environmental risks associated with transgenic crops. The overall objective will be to set an example/model highlighting India's achievement in conserving ecologically sensitive area such as mangroves, coral reefs, wetlands and representative ecosystems through biosphere reserves.

J. *Awareness, education, training and research*: Comprehensive packages are proposed to be developed on critical issues making them effective so that messages reach citizens like the pulse polio programme. Schools and colleges are proposed to be covered for participation in environmental programmes. Onthe- job trainings are proposed. Sustainable relationships with R and D institutions and academia will be developed for programme formulation and execution.

H. *Environmental law and policy*: The main theme will be revamping of environmental laws for ensuring their effective implementation, training of judicial officers in environmental laws and strengthening of enforcement mechanism.

Forestry

Forests are natural assets and provide a variety of benefits to the economy. Though the recorded forest area is about 23 per cent of the geographical area of the country much of this is degraded. There forests are therefore unable to play an important role in environmental sustainability and in meeting the forest produce needs of the people, industry and other sectors.

The problem and constraints in forestry development include lack of awareness about multiple roles and benefits of forests; especially its role in drought-proofing and prevention of soil and water runoff; no linkage between management and livelihood security of the people; low level of technology; inadequate research and extension; weak planning capability, wastage in harvesting an processing; market imperfections ; overemphasis on government involvement and control; low level of people's participation and NGOs involvement; lack of private sector participation; unwanted restrictions on felling, transport and marketing of forest produce grown by the people; lack of inter-sectoral coordination; and, conflicting roles of public forest administration. Existing policy, laws, rules, regulations and executive orders should be reviewed for removing constraints in holistic development of forestry with people's participation. Areas where action by government is needed and where the government programmes have not been up to the mark include:

— Focus on farm forestry which has been surprisingly diluted since 1991 despite its enormous potential, especially in agriculturally backward areas.

— Measures to sustain JFM beyond the project period have not been conceptualised. These relate to: building one-to-one correspondence between user groups and forest patch through a new forest settlement, recognisation of JFM groups in law and linking them with statutory panchayats, and integrating the activities of such groups with other income generating programmes such as watersheds and marketing of NTFPs.

— Protecting women's traditional usufruct rights and enabling women's group to collect and ma rket NTFPs has been proved to be viable and cost effective strategy not only for women's empowerment but also for renewal of forests. It is important to acknowledge and recognise women as managers, food gatherers, wage earners and producers. Forestry development policies must be developed accordingly.

— Continuing subsidies on government auctions of wood and bamboo to industries, which act as a disincentive to industry to pay a remunerative price to farmers. Governments need to examine the pattern of subsidy to forest based industries and wipe out that subsidy in a time-bound manner so as to improve valuation of forests.

— The tariff structure on forest-based products such as timber and pulp keeping in view the incentive effect on farmers.

— Poor understanding of the social implications of technology; old style plantations still continue to be funded despite the 1988 policy. Forest technology should be changed by shifting attention from timber to floor management and production of more gatherable biomass.

— Classification of bamboo as NTFP thereby providing rights to tribal population and other forest dwellers over this resource. Bamboo harvesting policy systematically maximises dry bamboo output for paper mills rather than green bamboo output for artisans. Infact the present practices ban felling of green bamboo. Streamlining the procedure for making green bamboo available directly to artisans.

— Gregarious flowering of bamboo is expected from 2003 onwards. This will result in mass mortality of bamboo, fire, increase of rodents, famines and unrest in tribal areas. Emergency plan should be formulated to face the calamity.

— Forest fire prevention and control should be given top priority to reduce losses and emission of greenhouse gases.

— Conservation and development of medicinal plants should be given priority to meet the local crude drug requirements and for export.

— Agroforestry, mountain, watershed development, river valleys, arid areas, wastelands afforestation programme should be given priority. · Research and technologic al development must increase productivity, production of new products, value addition, improved marketing, export and productive employment generation.

— Promotion of coastal shelterbelts plantations for prevention of natural calamities.

— An integrated system for addressing issues of animal welfare, including ecology, safety and commercial exploitation will need to be evolved.

Targets to bring greater areas under forest protection have been set for the five year periods 2002-2007 and 2007-2012. It has been proposed that the forest cover of the country should be increased to 25 per cent by 2007 and 33 per cent by 2012. This would mean bringing up extensive tracts of land under tree cover which would require a substantial investment both by the government and private sector.

The main elements of the strategy for 10th plan will include convergence of afforestation schemes to avoid duplicity and canalise smooth flow of funds; introduction of multi cropping system and promotion of agro forestry/ farm forestry; increased role and involvement of private sector in plantation; removing of bottlenecks and reviewing felling and transit rules and standardisation of cost norms under afforestation schemes. It also includes provision of quality planting material, infusion of improved and innovative technologies in nursery plantation, suitable protection against forest fires, pest attack and other diseases, strengthening of JFM net work for greater involvement of people and sustaining such programmes, strengthening of monitoring mechanism through establishment of IT network, development and sustainable management of NTFPs including medicinal plants, review of state monopoly on NTFPs, revival of working plans, strengthening of human resource planning and management and evolving Sustainable Forest Management (SFM) instrumentations.

Wildlife

Security of soil, sustenance of water regimes and activation of weather regimes as well as moderation of environmental adversities of modern age depend upon the health of natural ecosystems. The benefits from conservation of natural areas are not readily quantifiable as they flow to the local communities and the country somewhat intangibly and are often taken for granted. The value of effective conservation of natural areas must hence not be linked with direct fiscal benefits. Environmental well-being and food security cannot be ensured in the long run without the care of natural areas.

Efforts and initiatives from the side of the communities for revival of natural areas thus need to be supported. Challenges of wildlife conservation arise from population pressure, adverse impact of industrialisation as well as the lately increasing threats from illegal wildlife trade. Conservation of wildlife and its habitat outside the protected areas particularly in the context of critically endangered species is a challenge which the conservation efforts must address. Such efforts are equally needed for long ranging mega-herbivores like elephants and major predators like tiger and leopard. For the protection and conserving the wildlife, people's support has an important role.

Therefore bio-resource ameliorative and livelihood supplementing participatory ecodevelopment measures are required for local communities with a view to reducing their dependence upon protected areas. The problem of wildlife damage to human life and property need to be effectively mitigated by special management measures and expeditions and reasonable ex-gratia financial relief. Also, it is necessary to settle expeditiously the rights of local communities including voluntary relocation of habitations to alternative sites outside the protected areas. Such measures would ensure the participatory support of the people for conservation.

To fructify this objective, larger investments backed by attitudinal change would be essential. Poaching of wild animals due to lucrative clandestine trade in wildlife produce has become a major threat to wildlife conservation. Steps to be undertaken in this regard include filling up of vacant posts in the wildlife sector; reorganisation of field information in and around the protected areas; provision of infrastructure; capacity building; and, intelligence work. Education and awareness building measures as well as research and training have great role to play in making the conservation measures more effective.

In the ecosystem research scheme, it is proposed to take up studies relating to conservation, assessment of the impact of human activities on the biosphere, climate change and interrelationships between humans and biological resources. Under the Eastern and Western ghats scheme emphasis will be on location-specific and problem solving research. Efforts will be made to conserve, protect and enhance the biodiversity in Eastern and Western Ghats. Endeavors will be made for tackling the hotspots of pollution by development of biological and other interventions for prevention and control of pollution including waste recycling. Special attention is proposed to be given on human-animal interaction on the boundaries of wildlife sanctuaries, national parks, biosphere reserves and world heritage cites.

Attention will also be given to rehabilitation of mixed areas and over burdened dumps in the Eastern and Western ghats as also on strengthening of R&D on endemic fauna and climate change. The Ministry had identified about 21 thrust areas which will guide the selection of research projects during the Tenth Plan period. Some of the important areas of thrust are development of biological and other interventions for pollution preventing and control including waste recycling; development of strategies for prevention and/or of pollution; clean technologies for sustainable production patterns; restoration of degraded ecosystems; conservation and enhancement of biological diversity; climate change; health and toxicology; research on landuse and landuse changes; mining; impact of developmental activities on drinking water; river conservation and maintenance of their pristine state; research on mountain ecosystem, islands and coastal zones; research on environmental legislation; socio-economic issues.

5

India's Efforts for Environmental Protection

India was the first country to insert an amendment into its Constitution allowing the State to protect and improve the environment for safeguarding public health, forests and wild life. The 42nd amendment was adopted in 1976 and went into effect January 3, 1977. The language of the Directive Principles of State Policy (Article 47) requires not only a protectionist stance by the state but also compels the state to seek the improvement of polluted environments. This allows the government to impose restrictions on potentially harmful entities such as polluting industries.

An important subtlety of the directive's language is the provision that the article "shall not be enforceable by any court, but it shall be the duty of the State to apply these principles in making laws." This allows the directive to be an instrument of guidance for the government, while at the same time, since no law has been passed, no individual can violate existing law. Although state governments have clearly delineated lines of authority and jurisdiction, Article 253 of the Constitution provides the central government with sweeping powers to implement laws for any part of India with regard to treaties made with another country or decisions made by an international body.

For internal environmental matters, the constitution provides for a distribution of legislative powers between the union and the states. This was done by the creation of three jurisdictional listings; union, state and concurrent. The central government can enact a law on any items on the union and concurrent list, and in certain cases may enact a law on an item on the state list. To do so, the central government must approach the state legislatures. In the case of "water", which is an entry on the state list, Article 252 requires at least two or more state legislatures pass resolutions empowering the Parliament to pass water-related legislation.

Approaching the state on issues within its jurisdiction has historically led to cumbersome delays. For example, it was almost a decade after the first draft bill was being debated before the Parliament was able to pass India's first major water legislation in 1974. The process of state legislatures voting to empower the Parliament to pass water-related measures repeated itself during subsequent amendments to the water law in 1978 and 1988. The burgeoning awareness of the 1970's did not mark the first efforts by India on the environment.

In fact, the country has had a long history of environmentalism with the passage and codification of acts such as the Indian Penal Code, The Criminal Procedure Code, The Bengal Smoke Nuisance Act of 1905, The Indian Motor Vehicle Act, The Factories Act, The Indian Forest Act, The Mines and Minerals (Regulation and Development) Act, The Industries (Development and Regulation) Act, The Forest (Conservation) Act, The Merchant Shipping Act. The Indian Penal Code, passed in 1860, penalizes person(s) responsible for causing defilement of water of a public spring or reservoir with imprisonment or fines.

Traditionally, the interpretation of the Code has been viewed as a conservative attempt at enforcement. This is because punishment and fines have been characterized as meager. In addition, fouling a "public spring" has not, by definition, also included a "public river", which is where most pollution occurs. Finally, the specific language of the code places the burden of proof on the prosecution. Polluters must "voluntarily", "with intent", or "knowingly" discharge damaging effluents, making successful prosecution problematic in a court of law.

The Indian Forest Act was a product of British rule in 1927. The legislation granted the government uncontested rights over natural resources, with state governments authorized to grant licenses to lumber contractors and oversee protection of the forests. Even at this early stage, awareness of man's destructive tendencies was emerging. The Factory Act also addressed public safety and health issues. Section 12 of the Act empowered each state government to legislate its own rules and throughout the 1950's and 1960's individual states framed their own versions of the bill. The legislation addresses the discharge of water and effluents by factories, calling for effective arrangements for disposal at the plant-level.

As in the Indian Penal Code, penalties have been provided. As a result, a number of states passed versions of the Factory Act, including Uttar Pradesh in 1950, Tamil Nadu in 1950, West Bengal in 1958, Maharashtra in 1963 and Mysore in 1969. Each tailored the Act to suit its particular situation. In Uttar Pradesh, disposal of effluents had to have the approval of the state's Effluents Board. In Tamil Nadu, the ruling entity with similar responsibilities was the Director of Fisheries. In Maharashtra, local authorities were granted with jurisdiction in such matters.

The watershed event in the environmental movement was the Stockholm Conference on Human Environment in June, 1972. The conference made it apparent to all attendees that each nation needed to adopt comprehensive legislation addressing health and safety issues for people, flora and fauna. The United Nations, organizers of the conference, requested each participant to provide a country report. The findings by the Indian conferees shocked even the most pro-development advocates in India. Stockholm served as the genesis for the series of environmental measures India passed in the years to come. It has also been suggested that international events such as Stockholm provided the cover Indian officials needed to implement national environment policy without the vitriolic backlash normally expected from industry.

The Constitution permitted water-related issues to be controlled by the states, so prior regulatory efforts had been considered the domain of localities. In fact, during the 1950's and early 1960's, several states had taken steps on water protection. Laws passed included The Orissa River Pollution Act of 1953, The Punjab State Tubewell Act of 1954, West Bengal Notification No. 7 Regulation - Control of Water Pollution Act of 1957, Jammu and Kashmir State Canal and Drainage Act of 1963. The Maharashtra Water Pollution Prevention Act was passed in 1969.

As early as 1962, however, the Ministry of Health had begun to address water pollution issues by appointing a study committee. The committee made recommendations for both central and state level action. Jurisdictional questions remained unsolved between the states and central government, but by 1965 a draft bill was finally being circulated which allowed the states to pass resolutions authorizing Parliament to enact legislation on their behalves. By 1969, a bill, the Prevention of Water Pollution, had been introduced. Ultimately, a modified version, the Water (Prevention and Control of Pollution) Act, passed in 1974. Institutionalizing a regulatory agency for controlling water pollution marked the first true commitment on the issue by the Indian Parliament.

The Water Act also established the Pollution Control Boards at central government and state government levels. In the aftermath of the Water Act, Prime Minister Indira Gandhi moved to enact a series of environmental measures. The Department of Environment, created in 1980, performed an oversight role for the central government. DOE did environmental appraisals of development projects, monitored air and water quality, established an environmental information system, promoted research, and coordinated activities between federal, state and local governments. DOE was criticized, however, by environmental groups who recognized that with its small political and financial base the agency was weak and symbolic in nature.

Environmentalists recognized quickly that DOE would essentially serve as an advisory body with few enforcement powers. The government of Prime Minister Rajiv Gandhi recognized these deficiencies as well, and shortly after his administration began in 1985 created the Ministry of Environment and Forests (MoEF). MoEF was more comprehensive and institutionalized, and had a Union Minister and Minister of State, two political positions answering directly to the Prime Minister. The agency was comprised of 18 divisions, and two independent units, the Ganga Project Directorate and the National Mission on Wastelands Development. It continued the same functions that DOE originally had, such as monitoring and enforcement, conducting environmental assessments and surveys, but also did promotional work about the environment.

At its creation, MoEF had 684 employees. It had 18 new divisions, including the Division of Pollution Control. By 1989, it had 1,171 personnel and was still expanding. Its budget changed radically as well. In 1982-83, the agency had a total budget of Rs. 138.643 million, with Rs. 21.5 million designated for the Pollution Control Division. Its funding covered the Central Board's expenses, cess reimbursements to the state boards, the budget for the Clean Ganga Project, laboratories and other pollution control schemes.

By 1988-89, the overall budget stood at Rs. 18.2 million. And by 1984-85, the pollution control budget had jumped to Rs. 169 million, more than Rs. 30 million more than the entire budget for the agency two years earlier. The MoEF's implementation of a monitoring system was also aggressive. In 1977, India had 18 monitoring stations for water. By 1992, there were 480 water stations, including 51 from the Global Environmental Monitoring System (GEMS). In 1984, the country had 28 air monitoring stations in seven cities. By 1994, the National Ambient Air Quality Monitoring Program had 290 stations in 99 cities.

The central government expanded its environmental bureaucracy in an incremental fashion. Laws were often written in vague language, leaving room for interpretation within the bureaucracy. The policy the central government tried to emphasize was to extend existing legal and regulatory policies, rather than breaking new ground. The hope was that in this way the potential controversies which often surround environmental legislation would be defrayed. In 1981, the Air (Prevention and Control of Pollution) Act was passed, and in 1986, Parliament passed the Environment (Protection) Act, designed to act as umbrella legislation on the environment.

The responsibility entrusted to administer the new legislation also fell to the central and state pollution control boards. The landmark bills, mentioned above, are not the only remedies pollution control boards use to enforce their environmental powers. Due to deficiencies in the Water Act, some state boards have been successful in persuading the Magistrates to use sections of the Criminal Procedure

Code to control water pollution. The code vests powers in the Magistracy for removal of a "public nuisance".

During the intervening years, other laws have been passed to address specific issues, including The Wild Life Protection Act, and The Atomic Energy Act. In December, 1993, the MoEF completed its Environmental Action Plan to integrate environmental considerations into developmental strategies, which, among other priorities, included industrial pollution reduction. In principle, though, the MoEF was already actively pursuing pollution abatement and prevention policies. Among the strategies employed by the MoEF, a Bank mission reported, was the implementation of a "polluter pays" principle. MoEF imposed significant increases in its water cess, increasing it six-fold from early 1992.

Since then, the MoEF has decided to increase the water cess again since the charges were still well below the current effluent treatment costs. The MoEF also decided to shift from concentration to load-based standards. This would add to a polluter's costs and remove incentives to dilute effluents by adding water, and strengthen incentives for adoption of cleaner technologies. It also issued water consumption standards, which were an additional charge for excessive water use. Targeting small scale industries has been an important task as well since these facilities greatly add to the pollution load. The Ministry provides technical assistance and limited grants to promote central effluent treatment plants. It has also created industrial zones to encourage clusters of similar industries in order to help reduce the cost of providing utilities and environmental services.

Water Pollution Act of 1974

The legislation establishes both a Central Pollution Control Board, and State Pollution Control Boards for Assam,

Bihar, Gujarat, Haryana, Himachal Pradesh, Jammu and Kashmir, Karnataka, Kerala, Madhya Pradesh, Rajasthan, Tripura and West Bengal, as well as the Union Territories. Each board, Central or state, consists of a chairman and five members, with agriculture, fisheries and government-owned industry all having representation. Some of the main responsibilities of the Central Board, pursuant to promoting cleanliness and pollution abatement of streams and wells, include: coordinating activities of state boards and resolving disputes among them; providing technical assistance; conducting investigations; opening laboratories for analysis of samples; establishing fees for different types of sample testing; researching issues and problems; training personnel; conducting media and public awareness campaigns; collecting and disseminating data on water pollution; and working with state boards to set standards by stream or well.

The state boards have similar responsibilities, although they also play an important subsidiary role of doing plant-level inspections and monitoring, and advising the Central Board of problems and trends at the local level. Plants can be required to provide the state with information on their pollution control technologies, and the state may acquire effluent samples, which are admissible in court. State board members also have unfettered access to any plant site at any time. In situations where a state board believes immediate action is necessary, it has the authority to prevent further discharges, and can also apply to a Judicial Magistrate for a restraining order.

In the case of an emergency, state boards are empowered to take whatever measures they deem necessary. The legislation also sets out specific penalties (prison sentences and fines) for violations of the Act. For example, anyone destroying board property, preventing a board employee from performing his or her duties, knowingly providing false information to the board, tampering with monitoring devices installed by the board can be imprisoned up to three months, or fined as much as Rs. 10,000, or both. More serious violations of the law can incur stiffer penalties, some as high as seven years of imprisonment or Rs. 5,000 per day fines.

As an additional deterrent, if a person convicted of an offense under this Act commits a similar offense afterwards, the court can have the offender's name, violation and penalties published in newspapers, with the expense of the publication recoverable as a fine. For those who disagree with the validity of the government's samples, they have an option of appealing to the central government in writing. The central government has final authority over the Central Board or joint board, as does the state government with the state boards.

As already noted, standards can be made specific to a river or well. However, the state government may, in consultation with, or on the recommendation of the state board, provide exemption from a standard. To address new or unanticipated issues which are particular to a plant or industry, state boards are empowered to provide consent orders regarding new discharges or outlets. Any person establishing an industry or facility discharging water pollutants must receive a consent from the state.

In its relationship with the state boards, the Central Board serves an advisory role. It cannot force a state board to adopt uniform standards or prosecute cases the state board is reluctant to. Another interesting facet of the Act is Section 19. It empowers a state government to restrict its application to a particular area (within the state) by declaring it as a water pollution prevention and control area. The rationale was to concentrate limited financial resources on the most critical areas. Since the Act's inception, however, the exercise of this power by state governments has often, and not surprisingly, led to controversies.

In 1980, a group of writ petitions, filed by several textile and print and dye plants, challenged the constitutionality of the section before the Rajasthan High Court. The petitioners contended Section 19 provided the state government with the power of arbitrary discretion. In the end the court ruled against the petitioners, noting the government had met the conditions stated in the section and had simply fulfilled the requirements mandated by the Act. The primary source of funding for the state boards comes from the Central Board and state government. This has also become a source of controversy because the state boards were at the mercy of local political forces and circumstances, usually out of their control, which withheld or prevented adequate funding.

The Water (Prevention and Control of Pollution) Rules of 1975 established the terms and conditions of service for members of the Central Board. This included salaries and allowances. The bill defined the powers and duties of the chairman and member secretary, and it also established the board's authority to hire experts on a temporary basis, such as consulting engineers.

The Water (Preservation and Control of Pollution) CESS Act of 1977 provided the Central and state boards with the authority to levy and collect a tax on industries using water. The measure's jurisdiction included all states, with the exception of Jammu and Kashmir. The tax is calculated on the basis of how much water consumed. During the debate of the bill, at question was whether, in the absence of any specific mention of a tax on water consumption in either the state or concurrent lists, the Parliament could exercise its residual powers. The High Court ruled it could because "Parliament had the exclusive power to make any law with respect of any matter not mentioned in the State List."

The Air (Preservation and Control of Pollution) Act of 1981 was the next area of focus following the passage of the Water Act. The legislation designated the Central Board and state boards, which governed water pollution duties, to also be empowered with the same authority and administrative functions for the Air Act. In essentially all respects, the functions and authorizations exercised by central and state officials in the Water Act remain the same. The Central Board sets national ambient air standards. In addition, the state boards have the power to petition local magistrates to restrain polluters from exceeding specified standards. This legislation, however, does not supersede provisions of an earlier law, the Atomic Energy Act of 1962, which addressed radioactive air pollution.

In 1986, following the tragedy of Bhopal, Parliament was motivated to address all environmental pollutants. In about six months, with only casual resistance from powerful industry and business interests, the government enacted The Environment (Protection) Act. Under this measure, the central government has responsibility for deciding standards, restricting industrial sites, laying down procedures and safeguards for accident prevention and handling of hazardous waste, oversight of investigations and research on pollution issues, on-site inspections, establishment of laboratories, and collection and dissemination of information. Samples collected by central government officials can be admissible in court.

The Department of Environment, Forests and Wildlife, which is within the Ministry of Environment and Forestry, was designated as the lead agency for administration and enforcement. Plants do have some leeway when it comes to meeting emission or effluent standards issued by the Central or state board. They have one year to comply. Should a state board wish to specify a more stringent standard, or decrease the time period a plant or locality has to comply, it may do so only with a written request to, and consent from, the Central Board. The bill also sets standards on specific pollutants in specific industrial sectors.

The measure provides guidelines for location of industries and mining areas, for permitting and restricting industries in environmentally sensitive areas, coastal zone regulations and environmental impact assessments of development projects. Committees convened to conduct EIA's must have disciplines in eco-system and water resource management, air and water pollution control, flora and fauna conservation, land use planning, social sciences, ecology and environmental health. Public hearings are also pre-requisite for project clearance. The measure also delineates a system where a manufactured product can receive certification as environmentally friendly or compatible. In one interesting rule under the Act, the state of Maharashtra was allowed to prepare a regional plan for the ecologically sensitive area of Dahanu Taluka.

Within this area, industries or industrial activities could be permitted or restricted based on environmental risk. Industries in the green category could be considered for approval or rejection by state agencies without prior approval from the MoEF. Industries in the orange category could be permitted in Dahanu Taluka with a proper environmental assessment and adequate pollution control measures approved by the MoEF. Industries listed in the red category are not be permitted to operate in Dahanu Taluka. The law also promulgates rules on hazardous waste management and handling. The act defines the responsibilities of handlers, circumstances for granting authorization, conditions of disposal sites, rules for importing hazardous wastes, reporting of accidents, packaging and labeling requirements and an appeal process for potential handlers who have been denied authorization.

Rules were also promulgated on the manufacture, storage and import of hazardous or toxic chemicals, micro-organisms, genetically engineered organisms, or cells. For issues regarding micro-organisms and cells, biosafety and approval committees can be convened with experts in pertinent fields serving as chairmen, co-chairmen and members. The legislation had another interesting component. For the first time, private citizens were given the right to file cases against non-complying factories. A private citizen may file a complaint, however, only after giving notice of at least 60 days to the concerned authority of his/her intentions to file.

One of the chief criticisms of the Water Act was its prohibition of the public from taking litigation against polluters. As a result, individual state boards were simply overburdened. One Gujarat official indicated its board could only handle about five percent of complaints. After reviewing the above mentioned legislative efforts, it is important to note the Central or state boards do not have the legal authority to close a polluting facility. However, they do have enough tools to effectively close a plant violating environmental laws or standards. For example, under the Hazardous Wastes (Management and Handling) Rules, 1989, a state board can grant a company authorization to handle hazardous wastes.

The board also may deny or rescind such authorization, thus effectively prohibiting the same company from engaging in this activity. Under current law, the state boards have the authority to designate sections of their jurisdiction as "heavily polluted" areas. This allows a board to prevent new industrial activities, or expansion of existing industrial activities, in this area. One example is the Vapi section of Gujarat. In another scenario, the Environmental (Protection) Act allowed the state of Maharashtra to prohibit certain highly polluting industries from operating altogether in the ecologically sensitive area of Dahanu Taluka.

Finally, the most obvious tool available to the pollution control boards is to apply to the Judicial Magistrate to restrain or prevent facilities from continuing to violate existing pollution laws. Should the court agree with the board that a plant's activities are in violation of the law and issue a restraining order to prevent further violations, it is quite possible the plant could no longer continue to function and would have to close.

Case Law and Public Activism

The Public Liability Insurance Act of 1991 provided public liability insurance for persons injured by accidents with hazardous materials. The measure mandates that business owners operating with hazardous waste take out insurance policies. An Environmental Relief Fund was established and is maintained by industry operators. During the 1980's, public sentiment, and critically, the Indian Supreme Court, took an increasingly sympathetic stance toward environmental security.

The outcome was a series of important public interest litigation decisions. The Union Carbide disaster in Bhopal not withstanding, the Doon Valley Case in 1985 marked a clear watershed. The Court ordered the closure of eight sizable limestone quarries because the facilities had adversely affected local water springs and the health of nearby residents. The Court noted its decision would cause the quarries to be closed "but it is a price that has to be paid for protecting and safeguarding the right of the people to live in a healthy environment...".

In the Delhi Gas Leakage Case, the Court introduced important legal norms of "strict liability" and "polluter pays". The Sriram Foods and Fertiliser Industries, Ltd., was allowed to reopen its facilities after agreeing to several important preconditions. These included ownership and management taking personal responsibility for compensation for any injuries or deaths caused by escaping chlorine gases, and establishing a fund for any victims of future gas emissions who would seek compensation. In addition, the Court took an unusual step by taking an active role in the environmental debate by suggesting the set up of Environmental Courts.

The courts would be convened on a regional basis, consisting of experts in science and technology, and would serve as an information bank for cases of pollution. The Court took a similar track during the Ganga Pollution Case, ordering the closure of a cluster of tanneries, unless they established primary treatment facilities within a few months. The tanneries appealed on the basis they did not have the necessary funds to purchase and/or install pollution control equipment. The Court, however, ruled that financial capacities were to be deemed "irrelevant".

The Court also went beyond the strict focus of the offending tanneries to condemn the lax enforcement imposed by the municipalities and state pollution control boards who had jurisdiction along the river. In another litigation, the Distillary Pollution Case, the courts judged that top executives of polluting firms were responsible for implementing provisions of the Water Act, unless the person(s) concerned was able to prove the offense was committed without his/her knowledge or that he/she exercised all diligence to prevent the offense.

A promising development in recent years for India's new environmental activism is the Ganga Action Plan. Begun in 1985, the project's principle aim was to address municipal sewage by setting up treatment facilities. With backing from the highest levels of the central government, a Central Ganga Authority, housed within the Department of Environment, was authorized with oversight powers. The CGA is comprised of experts in waste treatment, public health engineering and agriculture. This coordinated effort marked a significant departure from traditional oversight efforts and is considered an evolutionary step forward for the country's regulatory and enforcement bodies.

The effort indicated that the government viewed this pollution issue as a regional problem, rather than a local one, since the river affected a wide and disparate population. Even here, however, the project was hampered by delays and ineffective management. In January, 1994, the central government introduced new restrictions and prohibitions on the expansion and/or modernization of any activity or new project unless an environmental clearance is granted. Projects must submit to an Environmental Impact Assessment and an Environmental Management Plan. The industry sector and the size of the project determine whether the MoEF or the state government has jurisdiction in assessing the EIA.

However, in some highly polluting sectors such as power plants, cement and petrochemicals, the MoEF generally clears the EIA's. In addition, state governments may issue more stringent restrictions for heavily polluted areas. One example is the Vapi region in Gujarat, where the state now has a total ban on new activities for the country's 17 most highly polluting sectors. Despite the regulatory framework and oversight authority of the Central and state boards, environmentalists have viewed enforcement as remarkably lax. Boards have been historically lenient towards the discharge of polluting effluents by allowing industries to procure the "consent" required by the Water Act.

When violations have become too obvious to ignore, the Central Board has moved sporadically, often lethargically, to launch prosecutions. State boards have also been lax. From 1974, when the Water Act became law, till 1988, about 1,600 prosecutions were started by state boards. Less than 300 had been decided. And some responsibility for failing to successfully prosecute polluting industries falls squarely to the boards themselves. For example, by 1988 the state of Bihar had not set up a laboratory for testing water and air samples and had also not received appropriate authorization from the Central Board to institute proceedings under the Water Cess Act.

In the past, the outcome of litigation often were court mandates to set up effluent treatment facilities. However, industries continue to lack incentives to set up treatment equipment, or operate equipment if it already is installed, because operating it has been more expensive than non-compliance. No jail sentences, on record, have been meted out, and fines have not been considered punitive. The unintended effect is to provide industries with a "license to pollute" or "pay and pollute". And finally, the significant drawback of current environmental laws is language that puts the burden of proof on the pollution control boards, rather than the polluters.

The Water Act states that to enact punitive measures, the pollution of a water body must be done "knowingly" rather than "negligently". To provide a context for the behavior of the central and state pollution control efforts, some have argued a possible explanation is the general philosophy of an environmental agency within a developing country. In many Western and developed economies, environmental agencies emphasize conservation since there is excessive use of existing resources. In developing countries, scarcity of natural resources is less a concern than misuse of them.

Most central governments commit themselves to the pressing needs for economic development. Therefore, environmental ministries in developing countries tend to support development while trying to also incorporate ecological concerns with general government policies. Public activism is also worth noting. Public protests have concentrated on two subjects: the local industry whose pollution has affected nearby residents; and deforestation by the lumber industry. Interestingly, the genesis of most demonstrations have come from rural residents. One early example was the protest of treatment policies by the Orient Paper Mills in Madhya Pradesh.

The mills' effluents were killing local fish and cattle populations. Ownership claimed its treatment facilities were working effectively, yet animal and fish populations continued to die. In the absence of government action, protests continued, until finally the plant agreed to install a water treatment plant. Problems continued still, but the public's fervor was eventually dissipated by time and the industry's ability to influence the government to adopt a laissez-faire approach to the issue. Industry stonewalling became symptomatic in similar cases.

A similar case, however, came to a different conclusion because of an interesting twist. In an event that denotes India's continuing skepticism of international influence, protesters at the Zuari Agro Chemicals of Goa were successful because the target was a foreign subsidiary, U.S. Steel. In the end, the company provided compensation to local residents and installed an additional waste treatment plant. Protesters against forest loggers have had success through a campaign called "hugging", where trees are literally hugged to prevent loggers from cutting them down. The Chipko Aandolan group initiated this type of protest in response to logging near the Himalayas.

Opponents have questioned the protesters' motivation, charging the concern is over economic exploitation rather than environmentalism. In either case, the results were effective as the government abolished the "contract system" in the early 1970's. Since then, however, the "contractors" have been replaced by forest development "cooperatives" who have continued to use extraction and exploitation as part of management policy. One observation which can be drawn from the public awareness burgeoning throughout the 1970's and 1980's is that the public perceived the government, not as stewards, protecting the environment, but as allies of industry's interests.

The results were often public demonstrations against state and central government officials. The role of NGO's has become another dimension in the evolution of public activism. In the early 1960's only a handful existed. By 1983 about 250 environmental NGO's had registered with the government representing a 600% increase. Their numbers and influence continue to grow into the 1990's.

The Pollution Control Boards

The Central Pollution Control Board

The Central Board serves as the national board, with oversight powers over the various state boards. It is also the ruling body for the Union Territories. The Board was legislated into existence in 1977, following the implementation of the Water Act. However, by 1987, only 20 of 25 states had ratified the original measure and constituted their own state boards. Funding from the central government has increased steadily since the Board's inception in 1977. By 1987, its budget had increased more than ten-fold from its original allocation of Rs. 2 million to Rs. 21.2 million. At that time, most of the Board's expenditures were targeted toward inter-state monitoring, basin and sub-basin studies, comprehensive industry documentation and air pollution control.

Administrative expenses accounted for Rs. 5.63 million in 1987-88. Litigation costs rose from Rs. 30,000 in 1984-85 to Rs. 89,000 in 1987-88. The Board may prosecute polluting industries under Section 33 (apprehension of pollution) and Section 44 (violation of conditions of consent order) in the Water Act. Through its first ten years in operation, the Board's efforts at litigation were lethargic as it filed only 181 cases, and eventually dropped charges in 16 of those. For most of those years, only a few cases were filed annually. 104 cases were filed in 1986-87, marking a change in the Board's enforcement strategy.

By the Board's own estimates, the task of effective water pollution control was daunting. A 1981 survey indicated only 27% of major industries had full or partial treatment facilities. Moreover, industries were surveyed only in states which had already passed the Water Act. A 1985 Board estimate indicated only 3% of the country's population had access to a sewerage treatment system. India ranks 84th in the world in terms of the percentage of population with access to safe water. A total of 350 million Indians do not have such access.

Several obstacles stand in the Board's way for more effective prosecution of violators. First, the Board essentially controls only the administration of the Union Territories. Of those, only Delhi has significant amounts of industry. The second problem was an administrative misstep by the Board. It would routinely provide consent orders to industries, without forcing plants to set up treatment or abatement facilities beforehand. The final problem came from the courts. When the Magistrates decided on a case, it was usually to rule a treatment facility must be put in place. However, that is essentially the mandate of any consent order.

As a result, the Board was at the mercy of the courts to force delinquent industries to comply with the law. The industries, in turn, bogged down the courts with a myriad of appeals and adjournments. In 1991, the Board began implementation of an coordinated action plan for industrial pollution control with the state boards. The Board selected 17 manufacturing sectors that had been highly polluting. They were cement, thermal power, distilleries, sugar, fertilizer, integrated iron and steel, oil refineries, pulp and paper, petrochemicals, pesticides, tanneries, basic drugs and pharmaceuticals, dye and dye intermediates, caustic soda, zinc, copper and aluminum smelters.

In addition to identifying critical manufacturing sectors, the Board went on to determine which geographical locations had been most affected by industrial pollution. It identified 13 extremely polluted waterways. They were the Sabarmati, Subernarekha, Godavari, Krishna, Indus and its tributaries, Sutlej, Ganga and its tributaries, Yamuna, Hindon, Chambal, Damodar, Gomati and Kali. Following consultations with the state boards, 22 critically polluted areas around the country were also identified. They were: Vapi in Gujarat; Singrauli in Uttar Pradesh; Korba, Ratlam, and Nagoda in Madhya Pradesh; Digboi in Assam; Talcher in Orissa; Bhadravati in Karnataka; Howrah in West Bengal; Dhanbad in Bihar; Pali and Jodhpur in Rajasthan; Manali and North Arcot in Tamil Nadu; Visakhapatnam and Pataucheru in Andhra Pradesh; Chembur in Maharashtra; Najafgarh in Delhi; Govindhgarh in Punjab; Udyog Mandal in Kerala; and Parwanoo and Kala Amb in Himachel Pradesh. All these sites and rivers were targeted for short-term emergency programs.

The Board also sought to identify polluters by size. Again working with their state counterparts, the Board also identified 1,551 large and medium sized units throughout India. About two-thirds of the plants were located in a few key states: Andhra Pradesh (173); Gujarat (177); Maharashtra (335); Tamil Nadu (119); and Uttar Pradesh (224). 77% were primarily water polluting, 15% were air polluting, and final 8% were both. 1,125 were found to be in compliance. 319 plants were found to be not in compliance. Of those 319 plants, 258 had begun operating before 1981. The remaining 107 plants were ultimately closed.

The effectiveness of the various state boards during their formative years has also been problematic. Their ability to be vigilant toward pollution control has been compromised by funding difficulties. Under the Water Cess Act (the Water Act of 1974 is silent on the issue), any fees levied by a state board are sent to the central government by way of The Consolidated Fund of India. The central government reimburses the funds to the state government, which is expected to pass it back to the state board. While state boards, collectively, have increased revenues through the water cess, from Rs. 30.59 million in 1983-84 to Rs. 54.49 million in 1986-87, there have been numerous complaints of state governments failing to adequately return those revenues back to the boards.

The resulting effect on enforcement and monitoring has been noticeable. By 1987-88, all the state boards had filed a total of only 1,602 cases for prosecution under the Water Act. Of these, 288 had been decided and 1,314 cases were still pending. Recent signs indicate more vigilance by government officials toward violators. In 1994, Indian courts closed almost 1,000 factories for pollution problems. In addition, the Supreme Court fined 15 plants, including some multi-nationals.

The central government has made progress in building common effluent treatment plants with the Bank's help. Tax breaks and subsidies can finance up to 25% of capital costs. 150 plants so far are participating. For its part, the Bank will make $160 million available through the Industrial Development Bank of India. Seven treatment plants are now set to go, and 20 more are expected to begin operations by the end of 1996.

The cost of India's water pollution problems has been staggering. In 1982, the National Engineering and Research Institute estimated 70% of the nation's water was polluted, and 73 million workdays at a cost of Rs. 60 million were lost to health-related illnesses. The Bank also has taken a more activist stance on pollution control in India. The Bank has earmarked $355 million for new sewers and upgrades in Bombay. One Bank consultant has developed an environmental strategy for the city, though there are no specific projects for implementation at this point.

Currently, the Bank is engaged in an extensive pollution prevention and abatement project, loaning about $300 million. The project has: an institutional component to strengthen selected state pollution control boards; an investment component to support subprojects by individual firms for pollution abatement, with a focus on waste minimization, and adoption of cleaner methods of production; and a technical assistance component to help establish a "clean technology institutional network" for the purpose of promoting, developing and transferring new environmental technologies.

The Gujarat Pollution Control Board (GPCB)

The GPCB started in October, 1974 with a staff of 25. By 1994, the board had 349 personnel, of which 39 were engineers and 78 were scientists and analysts, a headquarters, three regional offices, two sub-regional offices, and a laboratory at each facility. The board's annual income in 1993-94, including grants, was Rs. 37 million. Its expenditures were Rs. 28 million. The state government provided Rs. 14.5 million. Large industrial units are levied a cess monthly, while small facilities are levied on an annual basis. Cess collection improved dramatically, rising from Rs. 6 million in 1991-92 to Rs. 13 million in 1993-94.

Total arrears in cess collection through March, 1994 was Rs. 47 million. Most of this was due from Municipal Corporations and local bodies. The increase in cess collections improved with the influx of new lab facilities and equipment, which allowed the board to increase its monitoring efforts. In 1991-92, it took 4024 water samples and 857 stack monitorings. By 1993-94, it took 4674 water samples and 1790 stack monitorings. During this same period between 1991-92 and 1993-94, annual water consent applications increased from 362 to 501 and air consent applications rose from 343 to 428. During 1992-93, consents granted were 246 for water and 238 for air, while there were 97 water rejections and 56 air rejections.

During 1993-94, there were 501 applications for water consents, with 298 granted, 121 rejected. That same year, there were 428 applications for air consents, with 328 granted and 40 rejected. By early 1993, the board had granted a cumulative total of 4632 consents under the Water Act and 3213 consents under the Air Act. 8501 consents had been granted overall. 1397 consent applications for water were rejected, as were 349 consent applications for air, and there was an overall total of 1776 rejections. Cumulative water totals through March, 1994 were 6636 applications, 4930 consents granted, 1518 rejections, and 188 applications under scrutiny. The cumulative air totals were 4068 applications, 3541 consents granted, 389 rejections and 138 applications under scrutiny.

In accordance with Section 17 of the Water Act, the board also advises the state government and Ministry of Environment about the location of industries likely to pollute. The board examines the location and pollution control processes before issuing location clearances. During 1993-94, the board received 1187 applications, issued 865 clearances, and rejected 204 applications. By March, 1994, the cumulative totals were 11,640 applications received, 9,366 granted, 1,980 rejected and 294 under review. Industries generating hazardous waste must also obtain authorization from the board for collection, reception, treatment, transfer, storage or disposal.

During 1993-94, the board received 392 applications, granted 249 authorizations and rejected 78 applications. By March, 1994, cumulative totals were 1225 applications received, 930 authorizations granted, 200 applications rejected and 95 under review. From its start in late 1974 through late 1985, the GPCB had filed only 84 cases. By March, 1994, the updated numbers were 2460 cases filed, with 1915 water cases, 328 air cases, and 217 cases under the Criminal Procedure Code (CPC). 925 cases had been decided, of which, 594 were water, 149 were air, and 182 were under the CPC. 1535 cases were pending, of which 1321 were water, 179 were air, and 35 were filed using the CPC.

Despite a noticeable increase in enforcement and monitoring, in December, 1994, Mr. Dilip Biswas, head of the Central Board, publicly declared Gujarat, with its industrial belt between Ahmedabad and Vapi, the most polluted state in the country. And Biswas specifically criticized the state board for being too lenient with industries. In the aftermath, in May, 1995, a state court ordered 25 industrial units in Ahmedabad closed. A study by a court-appointed panel found untreated effluent from the facilities had affected the drinking water and health of villagers as far as 40 kilometers away.

In fact, the panel even found that most of the 786 polluting units in Ahmedabad had primary or advanced treatment facilities, but these facilities were not being operated because "treating effluents doesn't bring in profits". The panel's report criticized the state board and the Gujarat Industrial Development Corporation for its ineffectiveness in preventing pollution. Shortly after the court-order closure of units in Ahmedabad, the court served 168 more units in Vapi with an ultimatum. It ordered the facilities to comply with pollution control standards within two months or close.

The Maharashtra Pollution Control Board (MPCB)

In 1970, the MPCB became the first state pollution control

board in India to begin operating. By 1993, the MPCB had a staff of 445. There were 167 inspectors, 51 analytical staffers, and 227 administrative personnel. For fiscal year 1993-94, the state government proposed expanding its personnel by 268 positions to 717. The board has a central laboratory at New Bombay, six regional offices, 19 sub-regional offices, two regional labs and two mobile labs. The MPCB recently has been constructing regional laboratories in Aurangabad, Nagpur, Nashik and Pune. A World Bank mission in 1994 judged the board still had not proved its effectiveness in monitoring and enforcement since the additional staff had not yet joined and the regional labs were still under construction.

The mission noted, however, collection of the water cess had increased from Rs. 21 million in 1991-92 to Rs. 56 million in 1993-94. Similar consent and other fees increased from Rs. 7 million to Rs. 17.5 million during the same time span. The Ministry reported moderately different figures with regard to revenues. The water cess reimbursement for 1991-92 was Rs. 18.8 million. Ministry data indicated sharp increases in funding from the Central Board and consent fee revenues from 1988-89 onward, and fluctuating funding levels from the state government. For 1991-92, the only year fully reported, the board had Rs. 60.465 million in funds. Of this, the state government provided 27% of total budgetary support, the Central Board provided 23%, cess reimbursement accounted for 31% and consent fees amounted to the final 19%.

The data also indicates steady increases in expenditures for capital and administrative costs. Expenditures for salary and pension increases escalated from Rs. 21.538 million in 1990-91 to Rs. 26.509 million in 1992-93. Those expenditures are projected to be Rs. 28.48 million in 1993-94. Still, the MPCB had an operating surplus of Rs. 66.366 million by early 1992. Future outlays for capital expenditures through 1998-99 were projected at Rs. 5.5 million for applied research, Rs. 10 million for a Pollution Control Action Plan, and Rs. 39.5 million for a training center.

Annual projected costs for the ongoing research, action plan, and training center are estimated at Rs. 10 million. With adjustments for pay raises and pension increases, total board expenditures are expected to rise from Rs. 40 million in 1994-95 to Rs. 59 million by 1998-99. There are approximately 190,000 industries in the state, of which 4000 are considered large or medium scale units. These facilities discharge nearly 120,000 pollutants.

By November, 1993, the MPCB had issued 68,270 consents since starting operations. Consents granted under the Water Act were 47,944. The Air Act accounted for 16,174. 4,152 consents were granted under other environmental measures. The board issues 6000 to 7000 consents per year. Since 1988-89 this figure has remained fairly consistent. Compliance can be viewed in two ways. By late 1993, only 24,472 of 190,000 industries who had already received consents were considered in full compliance. However, compliance rates have improved markedly. In 1988-89, only 7,396 plants were considered in full compliance. Sample collections have also improved steadily.

In 1988-89, the board collected 5,287 samples. By 1992-93, the number had risen to 6,727. Staff inspects large scale industries on a monthly basis, medium size plants every three months, and smaller facilities twice a year. In 1992-93, MPCB personnel had conducted 6,560 inspections. In 1993-94, the MPCB labs had analyzed 4,457 water, 1,771 groundwater, 1,140 ambient air and 440 stack emission samples. As of December, 1993, the board had filed 183 cases. The board had won 37 convictions, lost 20 cases and had withdrawn one. The final 125 were pending.

The Tamil Nadu Pollution Control Board (TNPCB)

The TNPCB started operations in 1982 with 17 people. By 1994, it had a staff of 638, of which 147 were technical and 73 were scientific personnel. The board has five regional offices, 16 district offices, three advanced labs to analyze effluents, ambient air quality and stack emissions surveys, another three district labs to conduct pollutant analysis, and six mobile labs. A Bank mission noted that accounts of the last five years show the board is financially self-sufficient. Between 1989 and 1994, 60% to 70% of the board's revenues came from consent fees, 10% to 15% from analysis fees and the final 10% to 15% from water cess reimbursements.

In fact, the Ministry of Environment reported during this period that despite a steady decline in funding from the state government, overall funding levels multiplied spectacularly. This was due to dramatic increases in consent fee collections, as well as cess reimbursements. In 1988-89, total funds amounted to Rs. 13.02 million. In 1993-94, projected levels were for Rs. 102.753 million. The state government provided Rs. 8.7 million in 1988-89, but was only projected to provide Rs. 1 million for 1993-94. Capital and operating expenditures have been following similar trends. In 1990-91, total expenditures were Rs. 43.22 million.

By 1993-94, they were projected to be Rs. 130.396 million. Surplus revenues for 1993-94 are projected at about Rs. 52.8 million. From surplus revenues accumulated in previous years, and unused capital grants, the board believed the capital expenditures of Rs. 81.476 million it had projected for 1993-94 would be offset. The board had budgeted significant investments for this fiscal year, and estimated capital outlays over the following four years would amount to Rs. 10 million annually. The board also estimated staff pay and pensions would rise annually at a rate of Rs. 50 million during this period.

The state has about 14,000 industrial units, of which 2300 are considered large and medium scale In Tamil Nadu, 119 plants belong to the 17 polluting sectors the Central Board designated as its highest priorities. Of the 119 plants, nine were in non-compliance, and 11 units have been closed. The TNPCB has pursued a strategy of creating common effluent treatment plants for small scale industrial waste. There are 28 such plants in the state in various stages of completion. By the end of 1993, the board had issued 10,927 consents. Annual figures have fluctuated dramatically.

In 1989-90, there 584 water and 513 air consents issued. By 1991-92, the numbers had risen to 2,813 for water and 2,536 for air. However, the next year, there were only 713 water and 555 air consents issued. The board made significant inroads in its ability to monitor plants. The number of samples analyzed by its labs increased from 4,145 in 1988-89 to 10,749 in 1992-93. During a five month stretch in 1993, 87 inspectors monitored 7772 industrial units and took 3194 emissions and effluent samples. The TNPCB had filed 311 cases under the Water Act, and 131 cases under the Air Act, for a total of 442 cases by the end of 1993. 187 cases were still pending. Successful prosecution rates were not provided.

The Uttar Pradesh Pollution Control Board (UPPCB)

The UPPCB has 450 personnel, of which 85 are inspectors and 43 are analytical staff. Its central HQ is at Lucknow, and there are 11 regional offices and 13 labs. The regional labs, at a cost of Rs. 28 million for land acquisition and construction, were completed in late 1994. In fiscal year 1993-94, the board requested and was granted the authority to hire 90 additional scientific staffers when the labs were completed. The board is almost entirely self-sufficient, raising almost 90% of its revenues through water cess reimbursements and consent fees.

The UPPCB has identified 6,571 industries in the state (The Ministry of Industries has identified 121,817 industrial units). 918 or 14% of these industries have been classified as heavily polluting, 1645 or 25% are considered moderately polluting and 4008 or 61% are categorized as less polluting. How classifications were made was not elaborated on. Of the highly polluting units, 224 belonged to the large/medium scale, with 130, or 58%, in compliance with statutory regulations. State owned enterprises, most notably in the thermal power and cement sectors, were not in compliance. Of the 400 prosecutions that the board had pursued by late 1994, the Uttar Pradesh High Court decided in their favor 122 times and decided against them 14 times. India's Supreme Court received 32 cases, and ruled in favor of the board 19 of 21 times.

The balance of the cases are pending. In one celebrated case, the Supreme Court issued a notice which required 511 air polluting industries in the Taj Trapezium area to detail the status of their pollution control efforts. 241 units either did not respond, or did not make progress with installation of equipment sufficient for the court. All were closed. 144 facilities ultimately complied with the court's order and reopened. 97 remain closed.

West Bengal Pollution Control Board (WBPCB)

The WBPCB, with headquarters in Calcutta, began operations in September, 1974. Like all the boards, the WBPCB's costs have risen significantly since the late 1980's. In 1987-88, the board received Rs. 4,582,656 in central and state government funding, along with fees derived from consents and certifications. By 1994-95, the total was Rs. 14.646 million. Expenditures followed a similar pattern. In 1987-88, expenditures, including salaries and rents, were Rs. 3.818 million. By 1994-95, expenditures had risen to Rs. 9.513 million. A closer look at the figures, however, shows the board has surpluses about half the time between 1987-88 through 1994-95.

In 1994-95, for example, there was a surplus of Rs. 5.132 million. During this same period, cess accounts rose from Rs. 3.108 million in 1987-88 to Rs. 3.585 million in 1994-95. A new loan agreement with Japan's Overseas Economic Cooperation Fund was recently finalized. The loan of 1525 million yen, set to begin in 1995-96, will finance the construction of a head office and laboratory, and two regional offices and the procurement of additional monitoring equipment. In 1994-95, the board issued 323 water consents. In 1993-94, there were 310 water consents issued, 125 air consents, 75 certifications for industries wishing to start operations and ten sanctions.

The number of certifications issued by the board in 1994-95 represents a dramatic rise over previous years. From 1986-87 to 1992-93, the number of issuance's ranged from a high of 137 to a low of 104. Water and air consents had a predictable rise during this period. In 1986-87, there were 184 water and 51 air consents. In 1992-93, there were 478 water and 142 air consents. The board has also recorded that there were 212 large and medium scale industries with installed effluent treatment facilities, and 40 more in the process of constructing some. 30 large and medium facilities have installed air devices that meet standards, while 131 other plants have installed devices which are being monitored.

In addition, the board has directed 171 other sites to install air devices. Under a state plan, the board has set up 24 air monitoring stations. Through the sponsorship of the National Ambient Air Quality Monitoring network, the board established four more air monitoring stations in Howrah and two station at Hooghly. In 1994-95, the board analyzed 360 effluent samples, 97 stack samples, 3529 ambient air samples, and 236 rivers samples.

By 1993-94, the board had initiated 54 cases in lower courts under a variety of regulatory measures. Of those cases, there were four convictions, and two were dropped. The rest were pending. In 1993-94, the board issued status reports to the Indian Supreme Court for both water and air. The board's report said 197 of 242 industries were meeting effluent standards, 16 were still under review, and 12 more had already closed for various reasons. Of the 181 plants monitored for air emissions, 60 had installed adequate devices and the rest had been given a deadline by the court to meet standards.

6

Waste Management

Due to both the natural and man-made activities, disasters occur every where. A hazard is a rare or extreme event that adversely impacts human life, property, or activity. No human settlements are free from the risk of natural hazards; therefore it is vital that researchers and decision-makers have access to all available hazards information. E&DM classified the Hazards and Disasters into four categories viz., Natural events, Technological events, Man-made events and Region-wise events (Table 1). The adverse impacts caused due to the indiscriminate disposal of Hazardous Wastes (HWs) come under the category of Environmental Disasters. s

Table 1: Types of Hazards and Disasters

S.No.	Types of Hazards & Disasters	Events
1	Natural Events	Avalanches, Cyclones, Droughts, Dust/Sand storms, Earth Quakes, Epidemic Diseases, Famines, Floods, Heat waves, Hurricanes, Landslides/ Mud slides, Lightening storms, Tornadoes, Typhoons, Volcanic eruptions, Wild-land fires.

2	TechnologicalEvents	Environmental Disasters, Fire and Explosion, Accidents (Industrial, Nuclear, Radiological, Transportation), Dam failures.
3	Man-made Events	Bio-Terrorism, Spreading Chemical Agents, Mass Hysteria, Sabotage, Assassinations, Vandalism etc.
4	Region-wise Events	Events specific to a region such as India.

There is a growing concern all over the world for the safe disposal of HWs generated from anthropogenic sources. HWs can be classified into-

(i) Solid wastes

(ii) Liquid wastes

(iii) Gaseous wastes

(iv) Sludge wastes.

High Powered Committee (HPC) defines HW as any substance, whether in solid, liquid or gaseous form, which has no foreseeable use and which by reasons of any physical, chemical, reactive, toxic, flammable, explosive, corrosive, radioactive or infectious characteristics causes danger or is likely to cause danger to health or environment, whether alone or when in contact with other wastes or environment, and should be considered as such when generated, handled, stored, transported, treated and disposed off. This definition includes any product that releases hazardous substance at the end of its life, if indiscriminately disposed off.

The HW needs to be disposed off in a secured manner in view of their characteristic properties. Severe pollution of land, surface and ground water may occur if the options available for Hazardous Waste Management (HWM) are not being efficiently utilized by the waste generators. As per the ideal industrial siting criteria in

India, the industry should have enough land available within its premises for the treatment and disposal and or reuse/recycling of the wastes generated from it. However, very few industries in India own proper treatment and disposal facilities.

Mostly the large-scale industries and a few medium-scale industries, and none of the small-scale industries own the above facilities. Among the 1440 industries identified in the country with a high potential for pollution, 203 industries (i.e., ~14%) are recorded as not having adequate facilities to comply with the regulatory standards for treatment and disposal of wastes. Financial, administrative, and infrastructural facilities are some of the reasons attributed for the above limitations. It is interesting to note that, till 1997, there is no secured landfill facility available in the country to dispose of HW. Hazardous Wastes (HWs) are disposed off at Treatment, Storage and Disposal Facility (TSDF), a centralized location catering to the HW generated from the waste generators in the near vicinity.

The TSDF will help the small and medium scale industries generating HW in disposing their wastes efficiently. Literature is available on the site selection and other related criteria of TSDF. The site selection criteria for a TSDF depend upon Receptors and Pathways of likely waste movement, Waste characteristics and Waste management practices. The planning for HWM comprises of several aspects ranging from identification and quantification of HW to development and monitoring of TSDF.

The Government of India has promulgated the Hazardous Waste (Management & Handling) Rules [HW (M&H)] in 1989 through the Ministry of Environment and Forests (MOEF) under the aegis of Environment (Protection) Act [E(P) Act], 1986. Under the HW (M&H) Rules, the hazardous wastes are divided into 18 categories. The details are given in Table 2. Moreover, the role and responsibilities of the waste generator, state/central pollution controls boards and state Government is clearly defined. In order to encourage the effective implementation of these rules, the MOEF has further brought out the Guidelines for HW (M & H) Rules in 1991 giving the technical details of the principles of HWM covered under the HW (M&H) Rules, 1989.

However, the selection of a suitable site for an effective functioning of TSDF is the key aspect and depends upon several factors such as waste characteristics, site characteristics, public acceptance and prevailing laws & regulations. The facility siting should also incorporate the protection of human health, environment and property values in a community. Though the selection of an ideal site confirming with the above factors is a difficult task, few Guidelines are available in India for selection of best site for the same purpose. In India, unauthorized dumping of HWs is however continuing and in most of the places, the HW is being utilized to fill the low-lying areas, which is not acceptable.

Table 2: Categories of Hazardous Wastes

Waste Category	Waste Type	Regulatory Quantities
1	Cyanide wastes	1 kg per year as cyanide
2	Metal finishing wastes	10 kg per year the sum of the specified substance calculated as pure metal
3	Waste containing water soluble chemical compounds of lead, copper, zinc, chromium, nickel, selenium, barium and antimony	10 kg per year the sum of the specified substance calculated as pure metal

4	Mercury, arsenic, thallium, and cadmium bearing wastes	5 kg per year the sum of the specified substance as pure metal
5	Non-halogenated hydrocarbons including solvents	200 kg per year calculated as nonhalogenated hydrocarbons
6	Halogenated hydrocarbons including solvents	50 kg per year calculated as halogenated hydrocarbons
7	Wastes from paints, pigments, glue, varnish, and printing ink	250 kg per year calculated as oil or oil emulsions
8	Wastes from Dyes and dye intermediates containing inorganic chemical compounds	200 kg per year calculated as inorganic chemicals
9	Wastes from Dyes and dye intermediates containing organic chemical compounds	50 kg per year calculated as organic chemicals
10	Waste oil and oil emulsions	1000 kg per year calculated as oil or oil emulsions
11	Tarry wastes from refining and tar residues from distillation or pyrolytic treatment	200 kg per year calculated as tar
12	Sludges arising from treatment of wastewater containing heavy metals, toxic organics, oils, emulsions, and spent chemicals, incineration ash	Irrespective of any quantity

13	Phenols	5 kg per year calculated as phenols
14	Asbestos	200 kg per year calculated as asbestos
15	Wastes from manufacturing of pesticides and herbicides and residues from pesticides and herbicides formulation units	5 kg per year calculated as pesticides and their intermediate products
16	Acid/alkali/slurry wastes	200 kg per year calculated as acids/alkalies
17	Off-specification and discarded products	Irrespective of any quantity
18	Discarded containers and container liners of hazardous and toxic wastes	Irrespective of any quantity

The essential elements of CHWM (Comprehensive HWM) are:

— *Identification of Hazardous Waste Generation*: Identifying the HW generating industries is the first step. The HWs are classified under 18 categories and this information (Table 2) may be used to screen the wastes generated and classifying them as HWs. However there are few observations that- there is a probability of occurrence of wastes in more than one category; and the above classification system does not give any information to understand the toxic characteristics of HW. Few suggestions are also given to improve the classification system. The data available with the State Industrial Development Corporation (IDC), District Industries Centre (DIC), State Pollution Control Boards etc. may be utilized to identify the industries with a potential for HW generation.

— *Data Collection*: After identifying the HW generating sources, the inventory of the data pertaining to HW generation can be developed by conducting surveys through specially prepared questionnaires to each of the identified sources. This should be followed by field visits for data verification. It is essential that, the data that is obtained from the above options is verified from secondary data (either published data or available for another industry producing similar products). HW generation rates estimated from Plant capacity in Indian industries are available in literature. This will help in identifying any misappropriate data and correcting in the database.

— *Waste Characterization*: The HW that is generated from the study region should be characterized. For this purpose, it is advisable that the samples may be collected from the waste generation source and analyzed in the laboratory. Literature data may be used in the absence of primary data. Detailed information on HW characterization pertaining to physical, chemical, and general characteristics; and properties pertaining to ignitability, corrosivity, reactivity, & toxicity is given by Babu and Gupta.

— *Quantification of Hazardous Wastes*: The HWs are quantified based on their individual characteristics. The several options of compatibility of wastes with different characteristics should be studied and segregated. The quantity of HWs will be expressed in terms of each category for disposal (e.g. Recyclable, Incinerable, or Disposable etc). The wastes that are recyclable are used/waste oil, lead wastes, zinc wastes.

— *Identification of sites for disposal*: After quantifying the HW, and assessing the probable area requirements for its treatment, storage and disposal, the sites are to be identified. For this purpose, toposheets and/or remote sensing images of the study region may be used. The sites are to be physically verified in the field and to draw observations pertaining to the four different types of attributes (viz., Receptor related-, Pathway related-, Waste characteristics related-, and Waste management practices related-) available for ranking the sites. The site with a minimum score out of the available sites for ranking should be chosen as the site for establishing TSDF. Physical models are available in literature that give accurate results than the approximate methods available in Guidelines for ranking the sites for TSDF.

— *Conducting EIA*: The Environmental Impact Assessment (EIA) should be conducted in the site identified in the above step. The impacts from the project should be identified and public acceptance should be obtained for clearing the site for TSDF. Of the available options for impact prediction, Matrix method gives a better approximation of results as it focuses on causecondition-effect relationship of the attributes involved in the activity. A typical matrix model with a four-scale ranking is available in literature that is applicable for TSDF projects.

— *Implementing TSDF Programme*: The TSDF programme should be implemented at the final designated site. The site should contain adequate provisions for storage, treatment (Stabilization, Incineration etc.) and final disposal. Layouts for collecting the HW from the waste generation sources should be planned. The site should have the laboratory facilities to monitor the function of TSDF from time to time. Landfill is the final disposal option in TSDF. The leachate that has percolated should be treated in Effluent Treatment Plant (ETP) before disposal. Monitoring of ambient environmental qualities and TSDF performance should be done regularly during the postclosure period of landfill (30 years).

Identification of Hazardous Waste Generation

The HW generation in Indian States is given in Table 3 (HPC, 2001; SDNP, 2003). The data shows that the HW generation is maximum in Maharashtra (45.47%) followed by Gujarat (9.73%). Minimum HW is reported in Chandigarh (0.0069%). The number of industries that generate HW are maximum in Maharashtra (30.38%) followed by Gujarat (22.93%). The data shows that, 13011 industries are generating 4415954 TPA of HW in India.

Table 3: Status of Hazardous Waste Generation in India

State / Union Territory	Code	Total Districts	Districts in which HW units located	Total units	Total HW generation TPA
Andhra Pradesh	AP	23	22	501	111098
Assam	ASS	23	8	18	166008
Bihar	BHR	55	12	42	26578
Chandigarh	CHN	1	1	47	305
Delhi	DEL	9	9	403	1000
Goa	GOA	2	2	25	8742
Gujarat	GUJ	24	24	2984	430030
Haryana	HAR	17	15	309	32559
Himachal Pradesh	HP	12	6	116	2159
Karnataka	KAR	27	25	454	103243
Kerala	KER	14	11	133	154722
Maharashtra	MAH	33	33	3953	2007846
Madhya Pradesh	MP	61	38	183	198669
Orissa	OR	30	17	163	341144
Jammu & Kashmir	JK	14	5	57	1221

Pondichery	PON	1	1	15	8893
Punjab	PUN	17	15	700	22745
Rajasthan	RJN	32	26	332	122307
Tamil Nadu	TN	29	29	1100	401073
Uttar Pradesh	UP	83	65	1036	145786
West Bengal	WB	17	9	440	129826
India	IND	524	373	13011	4415954

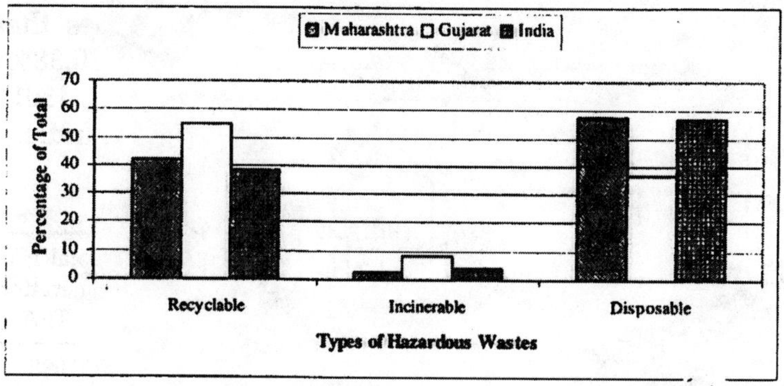

Figure 1: Comparison of Hazardous Waste characteristics in India

CHARACTERIZATION OF WASTE

The HW in India is characterized and documented in literature. The HWs are categorized into three groups viz., Recyclable, Incinerable, and Disposable. The details are given in Fig. 1 for Maharashtra, Gujarat and India (total). It can be noted from Fig.1 that, the HW generation trends in Maharashtra and India (total) are similar. The quantity of disposable HW (inorganic in nature to be disposed off in landfill) is high compared to the other two categories.

The quantity of HW generation reported in India is 4415954 TPA from 373 districts out of 524 districts (Table 3). According to one estimate, the land required to dispose

5.3 million tones of HW in an engineered landfill, assuming the average density of waste to be around 1.2 tonnes/m3 and the depth of the landfill 4 m, would be around 1.08 km2 every year. This information may be applied to future waste projections to arrive at future land requirements for the disposal of hazardous waste.

The number of sites identified for disposal of HW in India is 89 out of which 39 sites are notified. The State/ Union Territory wise status (HPC, 2001) of identification and notification of sites for disposal is given in Fig. 2.

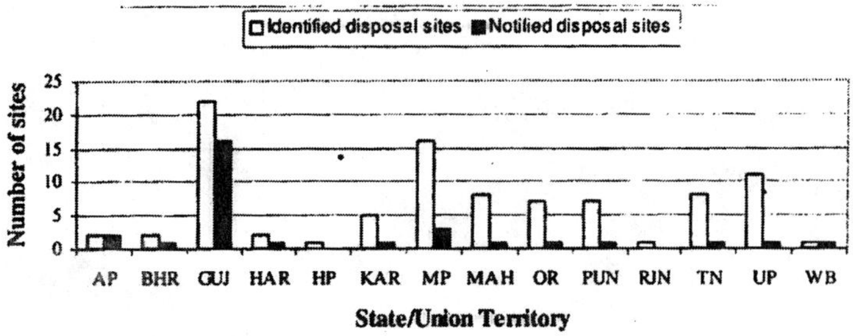

Figure 2: Status of Identified and Notified disposal sites in India

The Sensitivity Index (SI) for each attribute is evaluated on a four-level sensitivity scale ranging from 0 to 1 (0.0-0.25, 0.25-0.5, 0.5-0.75, 0.75-1.0). The aspects to be considered for attribute measurement are identified depending on the importance of the attribute. Based on the field data available, this attribute can be graded on the four-level-scale for the particular site. A total of 1000 points are divided among the four criteria of attributes @ 320, 280, 220, & 180 respectively using Delphi technique. Each of the 34 attributes is given weights based on the magnitude of its impact.

The value of the SI multiplied by the corresponding weightage would give the attributed score for each attribute. In the same way, score for all the attributes will be calculated and final attributed score for the site is obtained. This score is compared with the similar scores of the other sites available and all the sites are ranked as per the scores with the least score site given the top ranking. The total scores (out of 1000) can thus be interpreted in terms of the sensitivity of the site as follows:

— Score below 300: Very low sensitivity

— Score between 300 - 450: Low sensitivity

— Score between 450 - 600: Moderate sensitivity

— Score between 600-- 750: High sensitivity

— Score above 750: Very high sensitivity

A close examination of the options for ranking the sites has resulted in the following observations:

— The upper and lower limits for few attributes are not clearly defined.

— The sensitivity scale distribution for some of the selected attributes is not clear, and also nonlinear when overall distribution is considered.

— The error/ambiguity in the prediction of SSI could lead to erratic ranking of the site designated for TSDF.

Twelve attributes out of 34 attributes are identified having the above limitations. A model based on Regression analysis is developed to address the above limitations. The data given in Guidelines is taken as reference for the entire analysis.

— The analysis is carried out taking each attribute, case by case.

— Regression analysis is carried out to find out the Best-Fit Mathematical Model (BFMM) suitable for the data points of each attribute. Additional points are also generated by plotting graphs wherever necessary, for accurate fitting.

— An analysis is carried out by considering Linear Interpolation Among the Intervals (LIAI) (i.e., 0.25-0.5 & 0.5-0.75) specified in the Guidelines for all the data points.

— An additional analysis is also carried out for cross-checking by considering an Overall Linear Distribution Model (OLDM) of all the data points i.e., linear variation from 0.25-0.75.

— The above three analyses viz., BFMM, LIAI and OLDM are compared and conclusions are drawn.

The accuracy of the BFMM predictions is verified using Standard Deviation (SD) with respect to the expected value (i.e., 0.25, 0.5, & 0.75).

Conducting EIA

EIA serves as a valuable tool for identification, prediction and evaluation of impacts due to the proposed TSDF at a particular site. It evaluates the potential impacts, both beneficial and adverse, of the project i.e., TSDF on the environmental system. There are different methodologies available for the assessment of the impacts under the EIA study. However, the State Government or a person authorized by it will do the final selection of the site as per the Guidelines to HW (M&H) Rules issued by the MOEF (from time to time).

Evaluation of impacts is one of the important features of EIA. It summarizes and evaluates the impacts generated by taking up the TSDF. In view of the wide range of infrastructural and other associated requirements needed for site selection, construction and operation of TSDF, the impacts generated by it on the local environment become complex in nature. During the above stages of TSDF, the following phases are identified to take place:

— Construction phase

— Operational phase

— Final phase

The above phases may affect the local environmental attributes such as air, water, soil, land use, human beings and flora & fauna. Aspects such as access roads and services, site preparation, diversion of watercourses, infrastructural development, earth moving activities, traffic movements, leachate and gas control and/or treatment, re-vegetation, greenbelt development, monitoring etc. are addressed under the above activities. The local changes such as, public health and safety, population changes, changes in landscape, gaseous emissions, emission of water pollutants and local drainage, potential changes in local flora & fauna etc are also considered.

7

Forest Management

Centuries of exclusion disintegrated forestcommunity ties including the community system and practice of forest management. The mismatch between the growing needs of the communities for forest based requirements and the dwindling forest resources aggravated the pace of forest degradation.

Growing poverty in the countryside forced the local people to exploit the nearby forests for critical livelihood support by bringing forests under cultivation, selling small timber and other forest products, and even, in many cases, extracting the remaining rootstocks from forestlands.

Forest degradation had manifold impacts on the socio-economic life of the communities. The ecological fallout of such degraded surrounding forests negatively affected the local agriculture, animal husbandry practices and completely shattered the forest based livelihood of many. People started travelling to far off forest areas for need fulfillment resulting in conflicts with other communities and harassment by the forest department.

In such circumstances many communities gradually turned to their adjacent degraded forests and initiated protection measures perhaps as a last resort to restore back the forests and local livelihoods. Gradually, such local efforts turned the negative impacts of forest degradation into initiating factors for community-based forest management in the country. Local communities joined hands in bringing forestlands under their de facto control. Once protection by a few started, communities were quick to learn from each other and soon large tracts of forestland came under community protection and management.

Local villages protected even good forest areas in many parts of the country. The forest managing communities developed suitable forest management institutions and defined their membership. Elaborate rules and regulations were designed and operationalized to support the forest protection arrangement. Villages got into long drawn negotiations with other communities to define forest boundaries as well as the users who could be included as members.

Many communities from eastern India report that they approached the local forest department to seek permission for initiating protection activities in the close by degraded forest areas. However, their requests were turned down by forest officials who explained to them that the forests belonged to the department only and people did not have any rights to be involved in it, even if these forests were completely degraded.

Forced with dire consequences of degradation communities soon created institutions of forest protection and management and gradually took de facto control of such degraded forestlands. Forests which were brought under the de facto control of these local communities included Reserved, Protected and Revenue Forests. In most cases the forest department did not interfere because the lands were so degraded that one would doubt whether they would really regenerate.

Soon community efforts to regenerate degraded forestlands proved successful. Once the pace of degradation was arrested and the process of regeneration stabilized, forest benefits were quick to follow. Forest managing communities in various locations in India have reported that within a few years of successful protection they could derive forest benefits in terms of fuelwood, wood for agricultural implements, house repair and construction materials, forest foods and other non-timber products including fruits and leaves.

Two major developments were visible during this time: One, by their continued involvement in forest protection and with the beginning of benefit flow from forests, the communities created a far greater stake in the forests. Considerable strength was acquired by the forest management institutions. Their rule systems were established, membership formalized, negotiations on boundary and other local rights finalized with the neighbours, appropriation mechanisms for benefits sharing put in place and monitoring, sanctions and conflict management mechanisms evolved.

Having done their homework significantly well community forestry groups had realized that in order to sustain their forests and the forest management systems it is essential that the government devolves specific rights concerning forest and its tenure.

Forest Policy Environment

Two revolutionary forest policies during the late 1980's provided the base for the planners and administrators to be able to face the local realities in the forestry sector of the country.

The National Forest Policy of 1988 and the Central Government Guidelines for Joint Forest Management of 1990 made radical shifts from the previous forest policies, most specifically the National Forest Policy of 1952 which focused on forests for timber and stressing the need of industry and defense, and the recommendations of the National Commission on Agriculture 1976, which had approved commercial forestry to continue on forestland.

While creating a stable foundation for sustainable forest management in India the National Forest Policy of 1988 changed the focus of forest management from a highly "timber and revenue orientation" to ensure "environmental stability, maintenance of ecological balance and meet the subsistence requirements of the local people" by strengthening the people-forest link.

"The life of tribals and other poor living within and near forest revolves around forests. The rights and concessions enjoyed by them should be fully protected. Their domestic requirements of fuelwood, fodder, minor forest produce, and construction timber should be the first charge on the forest produce."

Advocating a stronger link between the people and the forests the Policy affirms: "Having regard to the symbiotic relationship between the tribal people and forests, a primary task of all agencies responsible for forest management, including the forest development corporations, should be to associate the tribal people closely in the protection, regeneration and development of forests as well as to provide gainful employment to people living in and around the forests.

While safeguarding the customary rights and interests of such people, forestry programmes should pay special attention to undertaking integrated area development programmes to meet the needs of the tribal economy in and around the forest areas, including the provision of alternative sources of domestic energy on a subsidized basis, to reduce pressure on the existing forest areas".

Joint Forest Management

Thus the 1988 National Forest Policy proposed the creation

of a people's movement to protect forest resources of the country. Following its footsteps in June 1990 the Central Government issued a circular on Joint Forest Management to operationalize participatory forest management and addressing the community forest protection activities in India.

The National Forest Policy, 1988, envisages people's involvement in the development of forests. The requirements of fuelwood, fodder and small timber such as house-building material, of the tribals and other villagers living in and near the forests, are to be treated as first charge on forest produce. The policy document envisages it as one of the essentials of forest management that the forest communities should be motivated to identify themselves with the development and protection of forests from which they derive benefits.

The salient features of the programme suggest by the Government of India resolution are:

— Access to forest lands and usufructory benefits will be given to those villagers who are organized into a village association;

— The beneficiaries are to be given usufructs like grasses, lops and tops of branches, and non-timber forest produce. On successful protection of forests, they are to be given a portion of the proceeds from the sale of trees when they mature;

— Along with trees with fuel, fodder and timber, village community may be permitted to plant fruit trees as well as shrubs, legumes and grasses;

— No grazing in the forestland protected by the village communities is to be allowed. But the villagers can cut and carry grass free of cost to promote stall feeding;

— No ownership or lease rights over the forest land including assignment of the forest land will be given to the beneficiaries;

— In case of failure to protect the area from grazing, encroachment etc., usufructory benefits could be withdrawn;

— The benefits of people's participation should be to the village communities alone and not to commercial or other interests.

In response to the national resolution of 1990, 22 state governments have come up with their respective notifications to operationalize the programme of JFM at state level. Though there are wide variations, Poffenberger and Singh, 1996 have found certain broad similarities in these state JFM resolutions.

— All resolutions provide user groups with usufruct rights only. They clearly note that land is not to be allocated or leased;

— The resolutions generally recommend village level committees as functional management groups;

— Village level groups are to operate under the supervision of forest department officials;

— In most cases the tenurial period for usufruct rights is not mentioned;

— Many resolutions limit community management and benefits to degraded forest land only;

— Some resolutions propose the formulation of joint management plans to coordinate agency and community activities. However, operational guidelines for cooperative planning are unclear, and where procedures are outlined, they tend to reflect forest department interests and concerns.

These resolutions are also silent about aspects of tenure on forestland. Rather, confusion persists regarding restriction of the micro plan period mostly to 5 years duration which is generally considered to be the period of tenure.

In the absence of clear tenurial system community forestry, even with the support of JFM (Joint Forest Management), is reeling under great uncertainties. This does not also help the government to develop an effective partnership with the village communities in the management of the forests. Moreover, this arrangement clearly gives the message of shortterm stakes and, as can be readily seen in many states, leads to people planning for short rotation production plantations Many state resolutions are prescriptive so far as the structure and functioning of the JFM institutions are concerned.

They suggest a uniform structure of JFM institutions by substituting the pre-existing-self-initiated traditional forest management arrangements; confining the unit of management to a revenue village or administrative boundary by breaking the historical links between the forest and several user groups; restricting the membership to individuals or households by undermining the value of universal adult membership; creating ex-officio posts in the village JFM institutions to be held by departmental representatives by excluding the community leadership; limiting the tenure of the institution to one or two years or till the end of project period (Tamil Nadu) by risking the continued existence of the institutions.

The effectiveness of the Self-Initiated institutions lies in their uniqueness rather than such uniform institutional arrangements. They represent immense variety in approach, methods and institutions that suit their local needs and specific problems. Having uniformity in the structure of the institution might help the forest department in better and simpler monitoring and reporting, but it may ruin the very spirit of these self initiated groups. It is the responsibility of the state to create adequate space for the community institutions to function independently.

Flexibility in policy framework to accommodate the existing forest management patterns is important. Each state JFM resolution provides for benefit sharing arrangement between the government and the village communities which is arbitrary in nature. JFM has largely been initiated on the degraded forest land and putting in a provision for timber or revenue sharing in such a preliminary stage seems to have defeated the purpose of long term forest management for ecological as well as local livelihood needs.

Because of the sharing provision, the communities look upon JFM as an encroachment by the Forest Department of forests which have been protected by the village communities even before the advent of JFM. The benefit distribution systems that these village communities have evolved over time largely pertain to the specific need based requirements. Final felling presupposes a one-time harvest and might again leave the land at square one. It has an inherent message of commercial production forestry, which is incompatible with village forests, or with sustained forest cover.

The fact that a massive programme like JFM is based on government orders puts the whole concept of participatory forest management in a disadvantageous position. At a time when management of forests by local people has gained momentum and the positive results of it are clearly visible, the legal aspects need careful and urgent consideration. As perceived by thousands of forest protecting communities, without a policy guard and statutory basis, JFM might remain an experimentation tool in the hands of the forest department and acquire a temporary character soon.

JFM resolutions of various states suffer from several operational shortcomings too. Issues pertaining to boundary demarcation, type of forest area to be included, registration of the forest protection committees, benefit sharing, role of various agencies including forest department and Panchayats etc. have remained problematic in the implementation of JFM. Moreover, JFM suffers much from the procedural maladies. This problem could be addressed at the Department level through internal monitoring mechanisms.

Forests cannot grow in isolation, so is true in case of its management. In order to attempt a suitable policy intervention for effective local management of forests, under the concept of joint management, there is a need to bring corresponding changes in related Acts and Policies. For example, JFM talks about 100% intermediary products to community but, at the same time, several restrictive Minor Forest Produce Acts and Policies, advocating monopoly leases and nationalization of Minor Forest Produce exists.

Similarly, while JFM gives primacy to micro plans there are no changes in the corresponding acts upholding the supremacy of the Working Plan. There have been enormous changes in the forest management scenario of the country over the last few decades. However, it is ironical that the same forest department, without changes and with almost the same orientation, remains the nodal agency still. Several states are currently experiencing the debate for organizational changes in forest department to increase the effectiveness in a completely changed scenario of forest management.

This includes both structural as well as attitudinal (approach) suitability. In one of the workshops in Orissa the forest officials emphasized the need for creating mechanisms for organizational learning within the FD.

Inspite of its good intentions the JFM approach has not been successful in meeting the expectations raised by the National Forest Policy, 1988. Many questions have been raised regarding the implications of these new pronouncements.

Repository of the Legacy

The Indian Forest Act, 1927 continues in the present day without any significant changes in its provisions. The act is focused on two broad aspects of forest management covering the classification, declaration and management of forests into Reserved, Protected and Village Forests and the functions of forest department. Forests being in the concurrent list, the states have also come up with State Forest Acts at different periods of time during the pre-1988 Forest Policy which are same as the National Act with minor modifications.

The provisions of the Indian Forest Act as well as the State Forest Acts are antithetical to the prescriptions of both the 1988 and 1990 policies. While conservation, livelihood and community involvement are the guiding principles of these two policies, the Forest Acts still represent the concept of reservation, centralized management and people's exclusion. None of the state governments has issued any directive revising its state forest policy after the radical overhaul of Forest Policy by the Government of India in 1988.

Another Central Law, known as the Forest Conservation Act, 1980, was brought in with a view to checking further degradation. This Act imposes restrictions on the de-reservation of forests or use of forestland for non-forest purposes. While this Act was instrumental in checking the pace of forest degradation during the initial years, further amendments have brought in loopholes by creating certain spaces through which de-reservation and conversion of forestland could be possible with prior approval of the Central Government or appropriate authority.

Consequently, in India thousands of hectares of forestland is de-reserved every year even in the presence of a National Act on forest conservation restricting such moves. However, in the name of Forest Conservation Act Governments have often restricted community action concerning management and conservation of forestland. Poffenberger and Singh, 1996 observe that the Forest Conservation Act was passed to emphasize the social and ecological importance of these critical resources.

The Act placed controls on logging, while also recognizing the needs of communities. However, it did little to reverse the historic swing towards bureaucratic control. Infact, environmental concerns often imposed further limitations on community rights.

External Environment

One more important aspect in analyzing the context for tenurial security is the external environment which has become active in shaping the national policies and forest management priorities in India. Till the end of the Fourth Five Year plan India used its internal resources for development projects including forestry. Massive deforestation leading to crisis of food, fodder and fuel, recommendations of various committees and commissions like the National Commission on Agriculture, 1976, growing international concern for environmental degradation forced the government to look at forest development in a project mode.

Several such forestry projects attracted foreign funding on a large scale by exposing the forestry sector to the influences of the external environment. The country has already passed through an active phase of Social Forestry before the advent of the National Forest Policy of 1988 and the 1990 JFM circular. However, results of the social forestry plantations have not been supportive to the growth of the natural forests as well as the demand for more tenurial rights by the forest protecting communities in India.

Saxena 1996 observed that the funding availability for forestlands became quite precarious during the Social Forestry phase. As state funds were reserved to meet the matching contributions required for external assistance for projects on non-forest lands, forestlands were starved of funds, with several adverse effects. The neglect of forestland hurt forest dwellers and tribals. Subsequently, the JFM approach also fell into the trap of project mode.

The concept of Participatory Forest Management lured the international funders and large forestry (JFM) projects with external assistance came into existence. Recently, many Indian states have completed the implementation of such forestry projects and the results have been found not very encouraging. Project appraisal reports and independent studies have shown that the forest managing rural communities in India have been marginalized further as a direct result of these large forestry projects.

The forest department which had weakened with the degradation of forests leading to loss of its interest in the depleted forests, suddenly grew financially and administratively stronger with the inflow of huge external funds. While Social Forestry had focused primarily on the non-forest lands, these externally supported Forestry Projects even targeted forestlands for commercial plantations by reducing the existing community efforts in conservation to a mere token. In the externally supported forestry projects, despite the stated goals of local participation and social development, many local people were excluded from the decision-making and tribal people suffered adverse impacts to their welfare.

There were complaints from these groups that they had only learned of the project once implementation began. The new village committees have failed to facilitate local participation in forest management and do not address the concerns of the indigenous families who have got major livelihood dependence on forests. This was contradictory to the objective of involving the local community in forest protection, regeneration of degraded forests and developing alternative income generation resources.

These projects often lacked baseline studies of indigenous rights and access to forest resources and most of them have failed in securing customary resource rights and instead accepted the existing national legislation and local regulations that curtail indigenous rights. Many have also criticized the projects for favouring business and state agencies and discriminating against forest dwellers. Despite directives regarding security of land tenure, no steps had been taken in this regard.

As a result, the projects have intensified the conflict between the communities and the State bureaucracy instead of working towards ensuring the survival of both forests and the forest dependent. The developments in the forestry sector of the neighbouring Kingdom of Nepal have much to offer for our policy makers. Nepal has brought in favourable forest legislation and its government has taken a bold decision to devolve some forestlands to forest user groups in village communities, as per their size and need, in perpetuity. While this experience of Nepal motivates the forest communities of India in their efforts for secure tenure, our forest policies have much more to gain from it.

Community Owned Forests

Van Panchayats in the Kumaon Kimalayas

The Van Panchayats are enduring Village Forest institutions in the Kumaon Himalayas, recognized under a specific State Act and surviving for a century now. The origin of Van Panchayats can be traced to the numerous land, village and forest settlements carried out during the British thereby restricting the unlimited rights of people over forests and creating formal institutional control over the local forests.

Within a span of seven years all the 'benap'2 lands, which had already been declared as Protected Forests by 1897, were acclaimed as Reserved Forests resulting in massive curtailment of peoples' rights. This also saw the strongest ever resistance from the people to reclaim their rights and concessions forcing the government to constitute a Forest Grievances Committee which recommended the creation of Van Panchayats for the management of forests.

The lesser Reserved Forests were dereserved as class I forest and transferred to the Revenue Department for declaration of Van Panchayats. Forest Panchayat Rules were brought in during 1931, which continues in the present day with amendments in 1976. One-third of the villagers, through a resolution of the Gram Sabha (Village Council), can propose the formation of a Van Panchayat. Inquiries are made to ascertain the willingness of villagers to form Van Panchayat, the rights of villagers and the neighbouring villages and other objections.

The Van Panchayat Committee is formed through elections where any one having a right to vote can participate. The boundary of the Van Panchayat is determined by survey and preparing a map of the forest area. The rights and ownership of the members of the Van Panchayat are recorded in a book known as "Khasra".

Thus, the "Khasra" defines the forest as well as the members included under a specific Van Panchayat. The Van Panchayat lands are located within the revenue boundary of the villages and all residents and others who possess rights in the forest can be right holders in the Van Panchayat institution.

Van Panchayats are free to make their own rules and regulations to protect and manage the forest, grazing, grass and fuelwood collection, ensure equitable distribution of forest produces to the right holders, prevent encroachment and incident of fire in Van Panchayat land, etc. It has the right to form its own byelaws of such rules. The Panchayat Rules stipulate the following powers for the Van Panchayat members, which is given below. This includes both the rules which are in operation now and rules which have been modified or repealed in the 1976 amendments.

Rights and Responsibilities of the Van Panchayat

— Panchayat forestland cannot be sold, mortgaged or subdivided.

— The products and proceeds from the sale of products of the Panchayat forest are to be used for the benefit of the community.

— The Panchayat is to prevent villagers from cultivating the Panchayat forestland.

— The Panchayat is to demarcate the forest area.

— The Panchayat is to maintain minutes of meetings and records of accounts and make decisions on regular meetings.

Rights and Powers of the Van Panchayat

— Rights and powers of the Panchayat are similar to those of forest officials.

— Rule breakers are fined.

— The Panchayat can file court cases against rule breakers.

— Fees may be levied from users for fodder, grazing, fuelwood and other forest materials.

— Grazing in the Panchayat forest can be regulated, and animals that are found in the forest in contravention of rules may be impounded.

— Cutting instruments used in contravention of Panchayat ruled may be confiscated.

— Users who break rules regularly may have their rights restricted or suspended.

— Guards may be appointed by the Van Panchayat to monitor and enforce rules.

Rule Enforcement

— All fines imposed by the Panchayat are treated as government dues and recoverable using similar procedures.

Allocation of Income

— All income from the sale of forest products is allocated to rightholders as assigned to the Van Panchayat.

— Income from the sale of forest products (such as timber, resin, minor forest produce) to non-rightholders was assigned to the Van Panchayat.

Elections

— Panchayat officials are elected for five years and the members have rights to elect the members.

The Van Panchayats have undergone considerable retrogression after the amendments in 1976 after which approvals and permissions from the Revenue as well as the Forest Department are required for every activity by the Van Panchayats. The powers of the councils, especially their enforcement authority, suffered a substantial decline in 1976. The overall framework of rules within which they could operate became far stricter .

The chances of further degeneration of these Village Forest Institutions loom large after the recently revised Van Panchayat Rules 2001 which negates the very existence of these institutions by making a forester the member secretary-cum-joint account holder in it and bringing them under the fold of Joint Forest Management. This is not only an attempt at modifying the basic structure and nature of the Van Panchayat institutions, in practice, it might mean a final blow bringing an end to the century old tradition of Van Panchayats in this country.

Gramya Jungles in Orissa

"Gramya jungle" is a form of village forest created under the Survey and Settlement Act 1962 of Orissa state and managed by the village as per the provisions of the Survey and Settlement Rules 1962. Under this act and rules, it was recognized that a village and its inhabitants need land to be used for different purposes in their day to day life.

While attempting to draw up a land use pattern of a village, the Survey and Settlement Rules provide for reservation of government land for specific purposes in rural areas which includes grazing land, village forest, cremation ground, future developmental needs and common use of a village including homestead land. The rules specify that in every surveyed village 10% of the effect area of the village shall be reserved for the purposes of "Gramya Jungle" (village forest) and for communal and developmental purposes including homestead for future (rule 3a).

In the unsurveyed villages, double the area reserved for pasturage, which is fixed at 1 acre for every inhabitant, shall be reserved for the same purpose (rule 3b). The

Gramya Jungle lands are located inside the village boundary and recorded in the rights of records (RoR) of the village. The rules authorize the village institution to manage these jungle lands as per the norms of the village. The rules specify that "lands recorded as "Gramya Jungle" will be available for use only as village forest (rule 10b)".

The village has complete rights over the produces from these jungle lands which are apportioned among the villagers as per rules of the village institution. It has also the rights to make provisions for protection and management of the Gramya Jungle. Every villager has rights to participate in the management of Gramya Jungles and system of rule making and enforcement.

The legal ownership of the Gramya Jungle lands rests with the revenue department. With the imposition of extreme restrictions by the government on forest areas outside village boundaries, the Gramya Jungles have been used profusely by the villagers over the years. Once these lands were depleted people shifted their focus to the outside forest areas that were also fast degrading. However, most communities have now initiated re-protection of these lands through the village institution.

The institution of Gramya Jungle has also been affected by some of the recent policy changes. The forest conservation act 1980 brought these forests, though belonging to the revenue category and included in the village boundary, under its purview since it is suffixed by the word "jungle". It required that the management of these lands be transferred to the forest department while the legal ownership remains with the revenue department.

Once these lands come under the management of the forest department the village institutions are replaced with Van Samrakshyan Samity (VSS) under JFM and the entire management focus changes as per the rules made by the government. While the village community loses control over these inside forest areas, once recognized under state laws, they are virtually left with no forests which they could manage independently for the basic requirements of fuelwood, grazing, small timber etc.

Once under JFM a large part of the produce would have to be shared with the department. It is considered a reverse trend in the process of transfer of tenurial rights over forestlands to the village communities. From a situation of independent management control by existing village institutions, Gramya Jungles are being brought under the influence of an alien institution like Van Samrakshyan Samity under JFM.

This would mean a move to bring the Gramya Jungles from the legally recognized control of the village communities to a more restrictive control of the forest department. There are several community management systems, like the Mundari Khuntkatti in Chotanagpur, Community Forest Management in Orissa, Sacred Groves and Village Forests in many parts of the country, prevalent in India which are gradually getting marginalized due to unfavourable policy implementation.

Secure Forest Temure: Basis Rudimentary

Tangible Forest Boundary

Boundary of the forest area must be properly demarcated and made discernible prior to assignment of tenure. A properly demarcated forest confers the finest sense of security to the forest community. Tenurial confusions emerge due to lack of clear sense of physical limits of a particular resource leading to unperceivable notions about its boundary.

Forest communities have entered into vicious conflicts, both among themselves as well as with the Forest Department, in situations where the forests have been left with improper demarcation. It is foremost in the course of achieving tenure that the forest be demarcated judiciously, which in itself forms a strong foundation for tenurial security.

This is true both in case of private and common property resource regimes as in the process of their acquisition considerable emphasis is laid on making it tangible by way of appropriate demarcation and boundary. A well demarcated forest area does not automatically entail a physical boundary around the resource. Experiences in community forestry indicate that instead of encircling the forest with physical barriers it is the manner in which the process of forest demarcation is achieved holds importance. This may include an exercise of forest area specification after estimating the historical and current relationships and interactions of the forest community vis-à-vis the forest resource.

Forest area specification would not only establish the physical limits of the resource; it would also quantify the area for future reference and record. A tangible forest boundary gives a clear sense to the forest community to distinguish their forest areas as against others. It allows the forest community to set its limits in terms of physical area within which they can make arrangements for protection, management and use.

The neighbours or outsiders are able to see clearly that the specific forest area is demarcated and, hence they need to restrict their interference and free access habits. While in community forestry this is generally seen as mutual respect, it also brings in an essential element of local recognition to the boundaries of forest communities. In a recently concluded study on conflicts in community forestry, Nayak and Singh observe that several forest managing communities in eastern India are faced with the single problem of multiple boundary conflicts from the beginning of forest protection.

The Joint Forest Management complicated matters further by restricting the unit of forest area to certain hectares. Thus, what flows in practice is division of forest areas into mathematically equal units before allotment to JFM institutions. This grossly underestimates the historical factors influencing people-forest relationship and creates a vicious trap for the village communities to descend into endless conflicts.

In contrast, most successful traditional forest management groups have been able to overcome this problem by defining the forest boundary through a process of long drawn negotiations with all other probable users finally either including or excluding them. What emerges from these local negotiations is a tangible forest boundary and those who identify themselves with this resource constitute a user unit.

8

Environmental Ethics

The root word of ethics is ethos, meaning "fundamental spiritual characteristics of a culture." This "culture" can be an organization or a discipline, so we might speak of the academic ethos or scientific ethos. The word ethic, derived from ethos, refers to a precept or a founding or practiced principle. Environmental ethics and environmental philosophy are newcomers as academic fields. For ages, ethics and philosophy have dealt exclusively with human relationship to each other.

Philosophy for the ancient Greeks encompassed what we call political science, natural science, sociology, and psychology today. It was about the way humans formed communities and states and about obligations of individuals to each other and to the state. Ethics was part of it. The natural sciences, which later divided into disciplines such as chemistry, physics, and biology, were all part of "natural philosophy." As philosophy grew more and more apart from science, it became extremely anthropocentric, dealing with human issues-cognitive, political, and logical. Ethics became separated as a field of inquiry into what we "ought" to do, mostly with respect to other humans. Early ethics stayed closely tied to the religious sphere.

The academic field we call environmental ethics is relatively recent. The earliest articulation in formal environmental ethics may be Aldo Leopold's "Land Ethic," a apart of his Sand County Almanac that is a timeless statement about the need for change in our ethic toward nature, and the difficulties in effecting change. "No important change in ethics was ever accomplished," he wrote, "without an internal change in our intellectual emphasis, loyalties, affections, and convictions. The proof that conservation has not yet touched these foundations of conduct lies in the fact that philosophy and religion have not yet heard of it."

Values and Cultures

The word "value" means worth. It also refers to an ethical precept on which we base our behavior. Values are shaped by the culture in which we live and by our experiences. However, there are values that are held high by most cultures. These include fairness and justice, compassion and charity, duties and rights, human species survival and human well-being. While values guide our behavior, there are many behaviors to which we grow accustomed because of the society, culture, and conditions in which we live.

We may not explicitly examine our environmental values, for example, when we decide whether to live close to or far away from work. Or, perhaps it is more correct to say that we think of more our economic or social environment, comfort, and convenience when we make this choice. Such decision making by large numbers of people has had many serious environmental impacts, such as air pollution from large commuting populations, deterioration of the built environment in cities, and problems of environmental inequities.

A tension-or even a strong conflict-can occur between individual values and societal ones. As larger centralized technical systems are used for convenience and become a part of everyday life, one might find it hard to keep up with living in conformity with one's individual values. The environmental impacts of suburban living noted above is an example of this. Economic and social conditions may cause large numbers of people to have to ignore or compromise their environmental values. On another level, the lack of knowledge about environmental impacts may lead to decisions that have cumulative negative impacts on the environment.

Peoples' Attitudes Towards the Environment

The humanistic and moralistic value types described by Kellert involve a bonding with nature and a faculty of altruism or kinship toward her. Religions have formed the basis of many of our attitudes and have contributed to our primary ethos towards nature. Religions that consider "man" as a supreme creation, the noblest, or "made in the image of God," have defined an environmental ethic that sees nature and indeed all of creation beneath man and different from man, rather than man as a part of nature.

Ancient religions and traditions, on the other hand, see nature as our guardians, and humans as part of nature. These ancient "religions" are mostly philosophies of life-wisdom about life and living on earth-which derive their tenets from considering man as part of nature. They accord to nature a power and a role as our benefactor and our sustenance, and in turn, prescribes our obligations towards nature and the consequent values and attitudes. Thus Hinduism and native American "religions" considered pantheistic in the western worldview accord a sacred quality to all of nature, often by defining each element of nature as a deity.

Judeo-Christian and other man-centered religions believe in a single, all-powerful God who made man in his image and gave him "dominion over the earth." Historian Lynn White's essay "The Historical Roots of our Ecological Crisis" attributes the modern exploitation of nature arising from the abuse of this "dominion," writing that "Christianity is the most anthropocentric religion the world has ever seen." This line of belief also engenders dualism, a separation between man and nature, even man versus nature. Dualism, separatism, and reductionism are all paradigms or working principles that have come under question as we begin to recognize more and more how intertwined and interdependent all entities and processes are.

Dualism and separatism have long been convenient constructs used by civilizations in justifying domination of one group over another. While our topic here is the environment, and the dualism that defines man and nature as separate, with one superior to the other, dualism has been the principle behind other types of domination-man over woman, one race over another, invaders over the invaded "native" or "indigenous" groups, humans over animals, even one religion over another. Dualism is used to separate the dominating group as the "real," "genuine," or normal, and the rest as the "Other."

Reductionism is the other line of thinking that has indirectly led to harm of the environment. The progress of science and technology has gone hand in hand, and even owed its rapid rate to this mode of thinking. Reductionism, or reducing a system into its components and treating the interactions later or ignoring them altogether, has been a result of scientific, analytical modes of thought. It has made possible our knowing large amounts of details bout individual pieces of systems, with little knowledge about how they fit together.

Even realms of knowledge have been reduced to disciplines as we found the advantages of specialization-that we can know more, and that we can use and exploit that knowledge more effectively. Thus what was "natural philosophy" to Aristotle became specialized as science and philosophy further divided into physics, chemistry, biology, and so forth. Philosophy in turn was divided into logic, analysis, hermeneutics and other divisions. Two types of schisms developed as a result. First, knowledge as represented by science (derived from the Greek "scientia," meaning "to know") separated from philosophy, the love of wisdom. Philosophy itself became more analytical and "scientific." Second, knowledge and the practice of getting at knowledge through reductionist methods, got separated from ethics, which provided for reflective practice of linking impacts and standards of behavior with the doing of science.

More recently, there has been an attempt at the understanding of interrelationships. Systems science, chaos theory, and ecology are all examples of integrative science that try to understand nature including humans and our technology as a coherent whole. Industrial ecology is a particularly relevant example of systems thinking where industry and economy are viewed as systems interacting with the environmental systems.

Many cultures that have existed for over thousands of years evolved a relationship with nature, or more correctly, were an integral part of nature so that "Mother Earth," or "Mother Nature" is held as sacred. The full recognition of the close relationship between us and the Earth, or even the "one-ness" of us and the Earth/nature is a hard concept to understand fully in the way we construct knowledge in our analytical, reductionist ways of thinking. It is therefore important to look at the way we construct knowledge itself, including our culture's emphasis on knowledge as something received only form the outside and then internalized.

Knowledge, especially science, is considered valid only when all people independent of their origins and cultural contexts see the same thing. What this definition of knowledge does is to select out the facts that are independent of relationships which come from social, cultural, and personal contexts. Such "objective," "empirical" knowledge considers the knower as active and the thing being studied as passive. This in the study of nature and environment, nature is an object of study rather than a giver of knowledge and wisdom. Knowledge is seen as a human possession, a human activity, dominated by human interests.

All animate life interacts, and, to a greater or lesser degree, affects the life and behavior of all other animate forms. The Chipewyan interact with all life in accordance with their understanding, and the animate universe responds. This is the description of that tradition by anthropologist Henry Sharp. This means that a person's interaction, language, and behavior towards the natural world is mindful, as our interactions with people we care about, and this mutual respect is the underpinning of the Chipewyan environmental ethics.

Indeed we cannot even call it an "environmental ethic": it is their fundamental way of being-in communion with nature. The reciprocal relationship gives knowledge and language as acts mindful of nature, and nature in turn helps shape the knowledge. The relationship involves a different, respectful style of inquiry than when "Man," seen as superior, studies nature, seen as a passive "other." This type of worldview brings about an awareness that is both a way of knowing, and learning and a way of behaving.

If we are mindful of the rocks, the water, the air as not only things that we use for sustenance, but as witnesses to evolution on Earth, as our larger partners in life, knowledge and thoughts about the environment take on a different meaning in our lives. This type of thinking is woven into the fabric of belief of Native American cultures as well as Eastern cultures such as Hinduism.

These are really cultures rather than religions, or theology, or philosophy. These "wisdom traditions" are practiced not like modern religions such as Christianity or Islam but are ways of creating, discovering, and communicating worlds of meaning through ordinary and common actions and behavior. This is why stones are important in these traditions. The stones encode worldviews and paint pictures of the practices.

Technological Responsibilities

Technology often begins with a near-term, and often shortsighted response to a human need or want, and while it addresses the immediate need, often causes long-term consequences that are deleterious. This is most evident in the state of the environment, and in the rise of chronic diseases such as cancer, and reproductive and immune dysfunction.

Modern technology, informed by an ever deeper penetration of nature and propelled by the forces of market and politics, has enhanced human power beyond anything known or even dreamed of before. It is a power over matter, over life on earth, and over man himself; and it keeps growing at an accelerating pace. Its unfettered exercise for about two centuries has raised the material estate of its wielders and main beneficiaries, the industrial 'West,' to heights equally unknown in the history of mankind. But lately, the other side of the triumphal advance has begun to show its face, disturbing the euphoria of success with threats that are as novel as its welcomed fruits...the peaceful and constructive use of worldwide technological power, a use in which all of us are collaborators as captive beneficiaries through rising production, consumption and sheer population growth - that poses threats much harder to counter. The net total of these threats is the overtaxing of nature, environmental and (perhaps) human as well. Thresholds may be reached in one direction or another, points of no return, where processes initiated by us will run away from us on their own momentum - and toward disaster.

As we understand the complex human need/want - industry - impact systems and cycles, we need to predict and avoid the consequences by wiser design and use of technology. The most eloquent voice that spoke about the effects of pesticides, one that was heard around the worlds, and mobilized the environmental conscience of the United States was that of Rachel Carson, a young biologist and a Pittsburgh native.

Environmental ethics stretches classical ethics to a breaking point. All ethics seeks an appropriate respect for life. But we do not just need a humanist ethic applied to the environment, analogously to the ways we have needed one for business, law, medicine, technology, international development, or nuclear disarmament. Respect for life demands an ethic concerned about human welfare, like the others and now concerning the environment. But environmental ethics in a deeper sense stands on a frontier, as radically theoretical as it is applied. Alone, it asks whether there can be nonhuman objects of duty.

Neither theory nor practice elsewhere needs values outside of human subjects, but environmental ethics must be more biologically objective—nonanthropocentric. It challenges the separation of science and ethics, trying to reform a science that finds nature value free and an ethics that assumes that only humans count morally. Environmental ethics seeks to escape relativism in ethics, to discover a way past culturally based ethics. However much our world views, ethics included, are embedded in our cultural heritages, and thereby theory-laden and value-laden, all of us know that a natural world exists apart from human cultures.

Humans interact with nature. Environmental ethics is the only ethics that breaks out of culture. It has to evaluate nature, both the nature that mixes with culture and wild nature, and to judge duty thereby. After environmental ethics, you will no longer be the humanist you once were. Environmental ethics requires risk. Environmental ethics explores poorly charted terrain, where it is easy to get lost. One must hazard the kind of insight that first looks like foolishness.

Some approach environmental ethics with a smile—expecting rights for rocks and chicken liberation, misplaced concern for chipmunks and daisies. Elsewhere, you think, ethicists deal with sober concerns: medical ethics, business ethics, justice in public affairs, questions of life and death, peace and war. But the questions here are no less serious: the degradation of the environment poses as great a threat to life as nuclear war, and a more probable tragedy.

Higher Animals

Logically and psychologically, the best and easiest breakthrough past the traditional boundaries of interhuman ethics is made confronting higher animals. Animals defend their lives; they have a good of their own, suffer pains and pleasures like ourselves. Human moral concern should at least cross over into the domain of animal experience. This boundary crossing is also dangerous because if made only psychologically and not biologically the would-be environmental ethicist may be too disoriented to travel further.

The promised environmental ethics will degenerate into a mammal ethics. We certainly need an ethic for animals, but that is only one level of concern in a comprehensive environmental ethics. One might expect classical ethics to have sifted well an ethics for animals. Our ancestors did not think about endangered species, ecosystems, acid rain, or the ozone layer, but they lived in closer association with wild and domestic animals than do we. Hunters track wounded deer, the rancher who lets his horses starve is prosecuted. Still, until recently, the scientific, humanistic centuries since the so-called Enlightenment have not been sensitive ones for animals, owing to the Cartesian legacy.

Animals were mindless, living matter; biology has been mechanistic. Even psychology, rather than defending animal experience, has been behaviorist. Philosophy has protested little, concerned to locate values in human experiences at the same time that it dis-spirited and devalued nature. Across several centuries of hard science and humanist ethics there has been little compassion for animals. The progress of science itself smeared the human-nonhuman boundary line. Animal anatomy, biochemistry, cognition, perception, experience, and behavior, and evolutionary history are kin to our own.

Animals have no immortal souls, but then persons may not either, or beings with souls may not be the only kind that count morally. Ethical progress further smeared the boundary. Sensual pleasures are a good thing, ethics should be egalitarian nonarbitrary, nondiscriminator y. There are ample scientific grounds that animals enjoy pleasures and suffer pains; and ethically no grounds to value these in humans and not in animals. So there has been a vigorous reassessment of human duties to sentient life. The world cheered in the fall of 1988 when humans rescued two whales from the winter ice.

The ethics by extension now seems too nondiscriminating; we are unable to separate an ethics for humans from an ethics for wildlife. To treat wild animals with compassion learned in culture does not appreciate their wildness. Man, said Socrates, is the political animal; humans maximally are what they are in culture, where the natural selection pressures (impressively productive in ecosystems) are relaxed without detriment to the species Homo sapiens, and indeed with great benefit to its member persons.

Wild animals cannot enter culture; they do not have that capacity. They cannot acquire language at sufficient levels to take part in culture; they cannot make their clothing, or build fires, much less read books or receive an education. Animals can, by human adoption, receive some of the protections of culture, which happens when we domesticate them, but neither pets nor food animals enter the culture that shelters them.

Worse, such cultural protection can work to their detriment; their wildness is made over into a human artifact as food or pet animal. A cow does not have the integrity of a deer, a poodle that of a wolf. Culture is a good thing for humans, often a bad thing for animals. Their biology and ecology—neither justice nor charity, nor rights nor welfare—provides the benchmark for an ethics.

Culture does make a relevant ethical difference, and environmental ethics has different criteria from interhuman ethics. Can they talk? and, Can they reason?, indicating cultural capacities, are relevant questions, not just, Can they suffer? "Equality" is a positive word in ethics, "discriminatory" a pejorative one. On the other hand, simplistic reduction is a failing in the philosophy of science and epistemology; to be "discriminating" is desirable in logic and value theory. Something about treating humans as equals with bighorns and cougars seems to "reduce" humans to merely animal levels of value, a "no more" counterpart in ethics of the "nothing but" fallacy often met in science.

Humans are "nothing but" naked apes. Something about treating sheep and cougars as the equals of humans seems to elevate them unnaturally, unable to value them for what they are. There is something insufficiently discriminating in such judgments—species blind in a bad sense, blind to the real differences between species, valuational differences that do count morally. To the contrary, a discriminating ethicist will insist on preserving the differing richness of valuational complexity, wherever found. Compassionate respect for life in its suffering is only part of the analysis.

Two tests of discrimination are pains and diet. It might be thought that pain is a bad thing, whether in nature or culture. Perhaps when dealing with humans in culture additional levels of value and utility must be protected by conferring rights that do not exist in the wilds, but meanwhile at least we should minimize animal suffering. That is indeed a worthy imperative in culture where animals are removed from nature and bred, but it may be misguided where animals remain in ecosystems. When the bighorn sheep of Yellowstone caught pinkeye, blinded, injured, and starving in result, 300 bighorns, over half the herd, perished.

Wildlife veterinarians wanted to treat the disease, as they would have in any domestic herd, and as they did with Colorado bighorns infected with an introduced lungworm, but the Yellowstone ethicists left them to suffer, seemingly not respecting their life. Had these ethicists no mercy? They knew rather that, while intrinsic pain is a bad thing whether in humans or in sheep, pain in ecosystems is instrumental pain, through which the sheep are naturally selected for a more satisfactory adaptive fit. Pain in a medically skilled culture is pointless, once the alarm to health is sounded, but pain operates functionally in bighorns in their niche, even after it becomes no longer in the interests of the pained individual.

To have interfered in the interests of the blinded sheep would have weakened the species. Even the question, Can they suffer? is not as simple as Bentham thought. What we ought to do depends on what is. The is of nature differs significantly from the is of culture, even when similar suffering is present in both. At this point some ethicists will insist that least in culture we can minimize animal pain, and that will constrain our diet. There is predation in nature; humans evolved as omnivores. But humans, the only moral animals, should refuse to participate in the meat eating phase of their ecology, just as they refuse to play the game merely by the rules of natural selection.

Humans do not look to the behavior of wild animals as an ethical guide in other matters (marriage, truth-telling, promise-keeping, justice, charity). Why should they justify their dietary habits by watching what animals do? But the difference is that these other matters are affairs of culture; these are person-to-person events, not events at all in spontaneous nature. By contrast, eating is omnipresent in wild nature; humans eat because they are in nature, not because they are in culture.

Eating animals is not an event between persons, but is a human-to-animal event; and the rules for this come from the ecosystems in which humans evolved and have no duty to remake. Humans, then, can model their dietary habits from their ecosystems, though they cannot and should not so model their interpersonal justice or charity. When eating they ought to minimize animal suffering, but they have no duty to revise trophic pyramids whether in nature or culture. The boundary between animals and humans has not been rubbed out after all; only what was a boundary line has been smeared into a boundary zone. We have discovered that animals count morally, though not yet solved the challenge of how to count them.

Animals enjoy psychological lives, subjective experiences, felt interests satisfied, intrinsic values that count morally when humans encounter them. But the pains, pleasures, interests, and welfare of individual animals are only one of the considerations in a more complex environmental ethic that cannot be reached by conferring rights on them or by a hedonist calculus, however far extended.

Organisms

If we are to respect all life, we have still another boundary to cross, from zoology to botany, from sentient to insentient life. In Yosemite National Park for almost a century humans entertained themselves by driving through a tunnel cut in a giant sequoia. Two decades ago the Wawona tree, weakened by the cut, blew down in a storm. People said: Cut us another drive-through sequoia. The Yosemite environmental ethic, deepening over the years, said: No! You ought not to mutilate majestic sequoias for amusement. Respect their life! Indeed, some ethicists count the value of redwoods so highly that they will spike redwoods, lest they be cut.

But trees and flowers cannot care, so why should we? We are not considering animals that are close kin, nor can they suffer or experience anything. Plants are not valuers with preferences that can be satisfied or frustrated. It seems odd to claim that plants need our sympathy, odd to ask that we should consider their point of view. They have no subjective life, only objective life. Perhaps the questions are wrong, because they are coming out of the old paradigm.

An organism is a spontaneous, self-maintaining system, sustaining and reproducing itself, executing its program, making a way through the world, checking against performance by means of responsive capacities with which to measure success. It can reckon with vicissitudes, opportunities, and adversities that the world presents. Something more than physical causes, even when less than sentience, is operating within every organism.

There is information superintending the causes, without it the organism would collapse into a sand heap. This information is a modern equivalent of what Aristotle called formal and final causes; it gives the organism a telos, "end," a kind of (nonfelt) goal. Organisms have ends, although not always ends-in-view. All this cargo is carried by the DNA, essentially a linguistic molecule. By a serial "reading" of the DNA, a polypeptide chain is synthesized, such that its sequential structure determines the bioform into which it will fold. Ever-lengthening chains (like ever-longer sentences), are organized into genes.

Diverse proteins, lipids, carbohydrates, enzymes—all the life structures are "written into" the genetic library. The DNA is thus a logical set, not less than a biological set, informed as well as formed. Organisms use a sort of symbolic logic, use these molecular shapes as symbols of life. The novel resourcefulness lies in the epistemic content conserved, developed, and thrown forward to make biological resources out of the physicochemical sources. This executive steering core is cybernetic—partly a special kind of cause and effect system, and partly something more: partly a historical information system discovering and evaluating ends so as to map and make a way through the world, partly a system of significances attached to operations, pursuits, resources. In this sense, the genome is a set of conservation molecules.

The genetic set is really a propositional set—to choose a provocative term—recalling how the Latin propositum is an assertion, a set task, a theme, a plan, a proposal, a project, as well as a cognitive statement. From this it is also a motivational set, unlike human books, since these life motifs are set to drive the movement from genotypic potential to phenotypic expression. Given a chance, these molecules seek organic self-expression. They thus proclaim a life way, and with this an organism, unlike an inert rock, claims the environment as source and sink, from which to abstract energy and materials and into which to excrete them. It "takes advantage" of its environment. Life thus arises out of earthen sources (as do rocks), but life turns back on its sources to make resources out of them (unlike rocks).

An acorn becomes an oak; the oak stands on its own. So far we have only description. We begin to pass to value when we recognize that the genetic set is a normative set; it distinguishes between what is and what ought to be. This does not mean that the organism is a moral system, for there are no moral agents in nature; but the organism is an axiological, evaluative system. So the oak grows, reproduces, repairs its wounds, and resists death. The physical state that the organism seeks, idealized in its programmatic form, is a valued state.

Value is present in this achievement. Vital seems a better word for it than biological. We are not dealing simply with another individual defending its solitary life but with an individual having situated fitness in an ecosystem. Still, we want to affirm that the living individual, taken as a "point experience" in the web of interconnected life, is per se an intrinsic value. A life is defended for what it is in itself, without necessary further contributory reference, although, given the structure of all ecosystems, such lives necessarily do have further contributory reference. The organism has something it is conserving, something for which it is standing: its life.

Organisms have their own standards, fit into their niche though they must. They promote their own realization, at the same time that they track an environment. They have a technique, a know-how. Every organism has a good-of-its-kind; it defends its own kind as a good kind. In that sense, as soon as one knows what a giant sequoia tree is, one knows the biological identity that is sought and conserved.

There seems no reason why such own-standing normative organisms are not morally significant. A moral agent deciding his or her behavior ought to take account of the consequences for other evaluative systems. Within the community of moral agents one has not merely to ask whether x is a normative system, but, since the norms are at personal option, to judge the norm. But within the biotic community organisms are amoral normative systems, and there are no cases where an organism seeks a good of its own that is morally reprehensible.

The distinction between having a good of its kind and being a good kind vanishes, so far as any faulting of the organism is concerned. To this extent, everything with a good of its kind is a good kind and thereby has intrinsic value. One might say that an organism is a bad organism if, during the course of pressing its normative expression, it upsets the ecosystem or causes widespread disease. Remember though, that an organism cannot be a good kind without situated environmental fitness. By natural selection the kind of goods to which it is genetically programmed must mesh with its ecosystemic role.

Despite the ecosystem as a perpetual contest of goods in dialectic and exchange, it is difficult to say that any organism is a bad kind in this instrumental sense either. The misfits are extinct, or soon will be. In spontaneous nature any species that preys upon, parasitizes, competes with, or crowds another will be a bad kind from the narrow perspective of its victim or competitor. But if we enlarge that perspective it typically becomes difficult to say that any species is a bad kind overall in the ecosystem.

An "enemy" may even be good for the "victimized" species, though harmful to individual members of it, as when predation keeps the deer herd healthy. Beyond this, the "bad kinds" typically play useful roles in population control, in symbiotic relationships, or in providing opportunities for other species. The Chlamydia microbe is a bad kind from the perspective of the bighorns, but when one thing dies, something else lives. After the pinkeye outbreak, the golden eagle population in Yellowstone flourished, preying on the bighorn carcasses. For them Chlamydia is a good kind instrumentally.

Some biologist-philosophers will say that, even though an organism evolves to have a situated environmental fitness, not all such situations are good arrangements; some can be clumsy or bad. True, the vicissitudes of historical evolution do sometimes result in ecological webs that are suboptimal solutions, within the biologically limited possibilities and powers of interacting organisms. Still, such systems have been selected over millennia for functional stability; and at least the burden of proof is on a human evaluator to say why any natural kind is a bad kind and ought not to call forth admiring respect.

Something may be a good kind intrinsically but a bad kind instrumentally in the system; these will be anomalous cases, however, with selection pressures against them. These claims about good kinds do not say that things are perfect kinds, or that there can be no better ones, only that natural kinds are good kinds until proven otherwise. In fact, what is almost invariably meant by a "bad" kind is that an organism is instrumentally bad when judged from the viewpoint of human interests, often with the further complication that human interests have disrupted natural systems. "Bad" as so used is an anthropocentric word; there is nothing at all biological or ecological about it, and so it has no force evaluating objective nature, however much humanist force it may sometimes have.

Species

Sensitivity to this wonder, however, can sometimes make

an environmental ethicist seem callous. On San Clemente Island, the U.S. Fish and Wildlife Service and the Natural Resources Office of the U.S. Navy planned to shoot two thousand feral goats to save three endangered plant species, Malacothamnus clementinus, Castilleja grisea, Delphinium kinkiense, of which the surviving individuals numbered only a few dozens. After a protest, some goats were trapped and relocated. But trapping all was impossible and many thousands were killed. We count plant species more than mammal lives, a few plants more than thousands of goats.

Those who wish to rare species of big cats to the wilds have asked about killing genetically inbred, inferior cats, presently held in zoos, in order to make space available for the cats needed to reconstruct and maintain a population genetically more likely to survive upon release. All the Siberian tigers in zoos in North America are descendants of seven animals; if these were replaced by others nearer to the wild type and with more genetic variability, the species could be saved in the wild. When we move to the level of species, we may kill individuals for the good of their kind.

Or we may now refuse to let nature take its course. The Yellowstone ethicists let the bison drown, callous to its suffering; they let the blinded bighorns die. But in the spring of 1984 a sow grizzly and her three cubs walked across the ice of Yellowstone Lake to Frank Island, two miles from shore. They stayed several days to feast on two elk carcasses, when the ice bridge melted. Soon afterward, they were starving on an island too small to support them. This time the Yellowstone ethicists promptly rescued the grizzlies and released them on the mainland, in order to protect an endangered species.

They were not rescuing individual bears so much as saving the species. Coloradans are considering whether to build the Two Forks Dam to supply urban Denver with water. This would require destroying a canyon and altering the Platte River flow, with many negative environmental consequences, including further endangering the whooping crane and endangering a butterfly, the Pawnee montane skipper.

Elsewhere in the state, water development threatens several fish species, including the humpback chub, which requires the turbulent spring runoff stopped by dams. Environmental ethics doubts whether the good of humans who wish more water for development, both for industry and for bluegrass lawns, warrants endangering species of cranes, butterflies, fish. A species exists; a species ought to exist. An environmental ethics must make these claims and move from biology to ethics with care. Species exist only instantiated in individuals, yet are as real as individual plants or animals.

The claim that there are specific forms of life historically maintained in their environments over time seems as certain as anything else we believe about the empirical world. At times biologists revise the theories and taxa with which they map these forms, but species are not so much like lines of latitude and longitude as like mountains and rivers, phenomena objectively there to be mapped. The edges of these natural kinds will sometimes be fuzzy, to some extent discretionary. One species will slide into another over evolutionary time. But it does not follow from the fact that speciation is sometimes in progress that species are merely made up, not found as evolutionary lines with identity in time as well as space.

A consideration of species is revealing and challenging because it offers a biologically based counterexample to the focus on individuals—typically sentient and usually persons--so characteristic in classical ethics. In an evolutionary ecosystem, it is not mere individuality that counts, but the species is also significant because it is a dynamic life form maintained over time.

The individual represents (re-presents) a species in each new generation. It is a token of a type, and the type is more important than the token. A species lacks moral agency, reflective self-awareness, sentience, or organic individuality.

The older, conservative ethic will be tempted to say that specific-level processes cannot count morally. Duties must attach to singular lives, most evidently those with a self, or some analogue to this. In an individual organism, the organs report to a center; the good of a whole is defended. The members of a species report to no center. A species has no self. It is not a bounded singular. There is no analogue to the nervous hookups or circulatory flows that characterize the organism. But singularity, centeredness, selfhood, individuality, are not the only processes to which duty attaches. A more radically conservative ethic knows that having a biological identity reasserted genetically over time is as true of the species as of the individual.

Identity need not attach solely to the centered organism; it can persist as a discrete pattern over time. Thinking this way, the life that the individual has is something passing through the individual as much as something it intrinsically possesses. The individual is subordinate to the species, not the other way around. The genetic set, in which is coded the telos, is as evidently the property of the species as of the individual through which it passes. A consideration of species strains any ethic fixed on individual organisms, much less on sentience or persons. But the result can be biologically sounder, though it revises what was formerly thought logically permissible or ethically binding. When ethics is informed by this kind of biology, it is appropriate to attach duty dynamically to the specific form of life.

The species line is the vital living system, the whole, of which individual organisms are the essential parts. The species too has its integrity, its individuality, its "right to life" (if we must use the rhetoric of rights); and it is more important to protect this vitality than to protect individual integrity. The right to life, biologically speaking, is an adaptive fit that is right for life, that survives over millennia, and this generates at least a presumption that species in niche are good right where they are, and therefore that it is right for humans to let them be, to let them evolve. Processes of value that we earlier found in an organic individual reappear at the specific level: defending a particular form of life, pursuing a pathway through the world, resisting death (extinction), regeneration maintaining a normative identity over time, creative resilience discovering survival skills.

It is as logical to say that the individual is the species' way of propagating itself as to say that the embryo or egg is the individual's way of propagating itself. The dignity resides in the dynamic form; the individual inherits this, exemplifies it, and passes it on. If, at the specific level, these processes are just as evident, or even more so, what prevents duties arising at that level? The appropriate survival unit is the appropriate level of moral concern. A shutdown of the life stream is the most destructive event possible. The wrong that humans are doing, or allowing to happen through carelessness, is stopping the historical vitality of life, the flow of natural kinds.

Every extinction is an incremental decay in this stopping life, no small thing. Every extinction is a kind of superkilling. It kills forms (species), beyond individuals. It kills "essences" beyond "existences," the "soul" as well as the "body." It kills collectively, not just distributively. It kills birth as well as death. Afterward nothing of that kind either lives or dies. "Ought species x to exist?" is a distributive increment in the collective question, "Ought life on Earth to exist?" Life on Earth cannot exist without its individuals, but a lost individual is always reproducible, a lost species is never reproducible. The answer to the species question is not always the same as the answer to the collective question, but, since life on Earth is an aggregate of many species, the two are sufficiently related that the burden of proof lies with those who wish deliberately to extinguish a species and simultaneously to care for life on Earth.

One form of life has never endangered so many others. Never before has this level of question—superkilling by a superkiller—been deliberately faced. Humans have more understanding than ever of the natural world they inhabit, of the speciating processes, more predictive power to foresee the intended and unintended results of their actions, and more power to reverse the undesirable consequences. The duties that such power and vision generate no longer attach simply to individuals or persons but are emerging duties to specific forms of life. What is ethically callous is the maelstrom of killing and insensitivity to forms of life and the sources producing them.

What is required is principled responsibility to the biospheric Earth. Human activities seem misfit in the system. Although humans are maximizing their own species interests, and in this respect behaving as does each of the other species, they do not have any adaptive fitness. They are not really fitting into the evolutionary processes of ongoing biological conservation and elaboration. Their cultures are not really dynamically stable in their ecosystems. Such behavior is therefore not right. Yet humanistic ethical systems limp when they try to prescribe right conduct here. They seem misfits in the roles most recently demanded of them.

If, in this world of uncertain moral convictions, it makes any sense to claim that one ought not to kill individuals, without justification, it makes more sense to claim that one ought not to superkill the species, without superjustification. Several billion years worth of creative toil, several million species of teeming life, have been handed over to the care of this late-coming species in which mind has flowered and morals have emerged. Ought not this sole moral species do something less self-interested than count all the produce of an evolutionary ecosystem as nothing but human resources? Such an attitude hardly seems biologically informed, much less ethically adequate. It is too provincial for intelligent humanity. Life on Earth is a many splendored thing; extinction dims its lustre. An ethics of respect for life is urgent at the level of species.

Ecosystems

A species is what it is where it is. No environmental ethics has found its way on Earth until it finds an ethic for the biotic communities in which all destinies are entwined. "A thing is right," urged Aldo Leopold "when it tends to preserve the integrity, stability, and beauty of the biotic community; it is wrong when it tends otherwise." Again, we have two parts to the ethic: first that ecosystems exist, both in the wild and in support of culture; secondly that ecosystems ought to exist, both for what they are in themselves and as modified by culture. Again, we must move with care from the biological claims to the ethical claims.

Giant forest fires raged over Yellowstone National park in the summer of 1988, consuming nearly a million acres, despite the efforts of a thousand firefighters. By far the largest fires ever known in the park, the fires seemed a disaster. But the Yellowstone land ethic enjoins: Let nature take its course. Let it burn! So the fires were not fought at first, but in midsummer national authorities overrode that policy and ordered the fires put out. Even then, weeks later, fires continued to burn, partly because they were too big to control, but partly too because Yellowstone personnel did not really want the fires put out.

Despite the evident destruction of trees, shrubs, and wildlife, they believe that fires are a good thing—even when the elk and bison leave the park in search of food and are shot by hunters. Fires reset succession, release nutrients, recycle materials, renew the biotic community. Aspen are important in the Yellowstone ecosystem. While some aspen stands are climax and self-renewing, many are seral and give way to conifers. Aspen groves support many birds and much wildlife, especially the beavers, whose activities maintain the riparian zones. Aspen are rejuvenated after fires, and the Yellowstone land ethic wants the aspen for its critical role in the biotic community.

Elk browse the young aspen stems. To a degree this is a good thing, since it gives elk critical nitrogen, but in excess it is a bad thing. The elk have no predators, since the wolves are gone, and as a result they overpopulate. Excess elk also destroy the willows and this in turn destroys the beavers. So, in addition to letting fires burn, rejuvenating the aspen might require managers to cull hundreds of elk—all for the sake of a healthy ecosystem. The Yellowstone ethic wishes to restore wolves to the greater Yellowstone ecosystem.

At the level of species, this is partly for what the wolf is in itself, but it is partly because the greater Yellowstone ecosystem does not have its full integrity, stability, and beauty without this majestic animal at the top of the trophic pyramid. Restoring the wolf as a top predator would mean suffering and death for many elk, but that would be a good thing for the aspen and willows, for the beavers and riparian habitat, with mixed benefits for the bighorns and mule deer, whose food the overpopulating elk consume, but who would also be consumed by the wolves. Restoration of wolves would be done over the protests of ranchers who worry about wolves eating their cattle; many of them also believe that the wolf is a bloodthirsty killer, a bad kind. Nevertheless, the Yellowstone ethic demands wolves, as it does fires, in appropriate respect for life in its ecosystem.

Letting nature take its ecosystemic course is why the Yellowstone ethic forbade rescuing the drowning bison, but rescued the sow grizzly with her cubs, the latter to insure that the big predators remain. After the bison drowned, coyotes and magpies, foxes and ravens fed on the carcass. Later, even a grizzly bear fed on it. All this is a good thing because the system cycles on. On that account rescuing the whales trapped in the winter ice seems less of a good thing, when we note that rescuers had to drive away polar bears that attempted to eat the dying whales. Classical, humanistic ethics finds ecosystems unfamiliar territory. It is difficult to get the biology right, and, superimposed on the biology, to get the ethics right.

Fortunately, it is often evident that human welfare depends on ecosystemic support, and in this sense all our legislation about clean air, clean water, soil conservation, national and state forest policy, pollution controls, renewable resources and so forth is concerned about ecosystem level processes. Further, humans find much of value preserving wild ecosystems and our wilderness and park system is impressive. Still, a comprehensive environmental ethics needs the best, naturalistic reasons, as well as the good, humanistic ones, for respecting ecosystems. Ecosystems generate and support life, keep selection pressures high, enrich situated fitness, evolve congruent kinds in their places with sufficient containment.

The ecologist finds that ecosystems are objectively satisfactory communities in the sense that organismic needs are sufficiently met for species long to survive, and the critical ethicist finds (in a subjective judgment matching the objective process) that such ecosystems are satisfactory communities to which to attach duty. Our concern must be for the fundamental unit of survival. An ecosystem, the conservative ethicist will say, is too low a level of organization to be respected intrinsically.

Ecosystems can seem little more than random, statistical processes. A forest can seem a loose collection of externally related parts, the collection of fauna and flora a jumble, hardly a community. The plants and animals within an ecosystem have needs, but their interplay can seem simply a matter of distribution and abundance, birth rates and death rates, population densities, parasitism and predation, dispersion, checks and balances, stochastic process. Much is not organic at all (rain, groundwater, rocks, soil particles, air), while some organic material is dead and decaying debris (fallen trees, scat, humus). These things have no organized needs. There is only catch-as-catch-can scrimmage for nutrients and energy, a game played with loaded dice, not really enough integrated process to call the whole a community.

Unlike higher animals, ecosystems have no experiences; they do not and cannot care. Unlike plants, an ecosystem has no organized center, no genome. It does not defend itself against injury or death. Unlike a species, there is no ongoing telos, no biological identity reinstantiated over time. The organismic parts are more complex than the community whole. More troublesome still, an ecosystem can seem a jungle where the fittest survive, a place of contest and conflict, beside which the organism is a model of cooperation.

In animals, the heart, liver, muscles and brain are tightly integrated, as are the leaves, cambium, and roots in plants. But the ecosystem community so-called is pushing and shoving between rivals, each aggrandizing itself, or else all indifference and haphazard juxtaposition, nothing to call forth our admiration. Environmental ethics must break through the boundary posted by disoriented ontological conservatives, who hold that only organisms are "real," actually existing as entities, whereas ecosystems are nominal--just interacting individuals.

Oak trees are real but forests are nothing but collections of trees. But any level is real if it shapes behavior on the level below it. Thus the cell is real because that pattern shapes the behavior of amino acids; the organism because that pattern coordinates the behavior of hearts and lungs. The biotic community is real because the niche shapes the morphology of the oak trees within it. Being real at the level of community only requires an organization that shapes the behavior of its members. The challenge is to find a clear model of community and to discover an ethics for it--better biology for better ethics.

Even before the rise of ecology, biologists began to conclude that the combative survival of the fittest distorts the truth. The more perceptive model is coaction in adapted fit. Predator and prey, parasite and host, grazer and grazed are contending forces in dynamic process where the well-being of each is bound up with the other--coordinated (orders that couple together) as much as heart and liver are coordinated organically. The ecosystem supplies the coordinates through which each organism moves, outside which the species cannot really be located.

The community connections are looser than the organism's internal interconnections--but not less significant. Admiring organic unity in organisms and stumbling over environmental looseness is like valuing mountains and despising valleys. The matrix the organism requires in order to survive is the open, pluralistic ecology. Internal complexity--heart, liver, muscles, brain--arises as a way of dealing with a complex, tricky environment. The skin-out processes are not just the support, they are the subtle source of the skin-in processes. In the complete picture, the outside is as vital as the inside. Had there been either simplicity or lock-step concentrated unity in the environment, no organismic unity could have evolved. Nor would it remain. There would be less elegance in life.

To look at one level for what is appropriate at another makes a category mistake. One should not look for a single center or program in ecosystems, much less for subjective experiences. Instead, one should look for a matrix, for interconnections between centers (individual plants and animals, dynamic lines of speciation), for creative stimulus and open-ended potential. Everything will be connected to many other things, sometimes by obligate associations, more often by partial and pliable dependencies and, among other things, there will be no significant interactions. There will be functions in a communal sense: shunts and criss-crossing pathways, cybernetic subsystems and feedback loops.

An order arises spontaneously and systematically when many self-concerned units jostle and seek their own programs, each doing their own thing and forced into informed interaction. An ecosystem is a productive, projective system. Organisms defend only their selves, with individuals defending their continuing survival and species increasing the numbers of kinds. But the evolutionary ecosystem spins a bigger story, limiting each kind, locking it into the welfare of others, promoting new arrivals, increasing kinds and the integration of kinds. Species increase their kind; but ecosystems increase kinds, superposing the latter increase onto the former.

Ecosystems are selective systems, as surely as organisms are selective systems. The natural selection comes out of the system and is imposed on the individual. The individual is programmed to make more of its kind, but more is going on systemically than that; the system is making more kinds. Communal processes--the competition between organisms, more or less probable events, plant and animal successions, speciation over historical time--generate an ever-richer community. Hence the evolutionary toil, elaborating and diversifying the biota, that once began with no species and results today in five million species, increasing over time the quality of lives in the upper rungs of the tropic pyramids. One celled organisms evolved into many-celled, highly integrated organisms.

Photosynthesis evolved and came to support locomotion--swimming, walking, running, flight. Stimulus-response mechanisms became complex instinctive acts. Warm-blooded animals followed cold-blooded ones. Complex nervous systems, conditioned behavior and learning emerged. Sentience appeared--sight, hearing, smell, tastes, pleasure, pain. Brains coupled with hands. Consciousness and self-consciousness arose. Culture was superposed on nature. These developments do not take place in all ecosystems or at every level. Microbes, plants, and lower animals remain, good of their kinds, and serving continuing roles, good for other kinds. The understories remain occupied.

As a result, the quantity of life and its diverse qualities continue--from protozoans to primates to people. There is a push-up, lock-up, ratchet effect that conserves the upstrokes and the outreaches. The later we go in time the more accelerated are the forms at the top of the tropic pyramids, the more elaborated are the multiple tropic pyramids of Earth. There are upward arrows over evolutionary time. The system is a game with loaded dice, but the loading is a prolife tendency, not mere stochastic process. Though there is no Nature in the singular, the system has a nature, a loading that pluralizes, putting natures into diverse kinds, nature1, nature2, nature3... naturen. It does so using random elements (in both organisms and communities), but this is a secret of its fertility, producing steadily intensified interdependencies and options.

An ecosystem has no head, but it has a "heading" for species diversification, support, and richness. Though not a superorganism, it is a kind of vital field. Instrumental value uses something as a means to an end; intrinsic value is worthwhile in itself. No warbler eats insects to become food for a falcon; the warbler defends its own life as an end in itself and makes more warblers as she can. A life is defended intrinsically, without further contributory reference. But neither of these traditional terms is satisfactory at the level of the ecosystem. Though it has value in itself, the system does not have any value for itself. Though a value producer, it is not a value owner.

We are no longer confronting instrumental value, as though the system were of value instrumentally as a fountain of life. Nor is the question one of intrinsic value, as though the system defended some unified form of life for itself. Ethical conservatives, in the humanist sense, will say that ecosystems are of value only because they contribute to human experiences.

Humans count enough to have the right to flourish there, but not so much that they have the right to degrade or shut down ecosystems, not at least without a burden of proof that there is an overriding cultural gain. Those who have traveled part way into environmental ethics will say that ecosystems are of value because they contribute to animal experiences or to organismic life. But the really conservative, radical view sees that the stability, integrity, and beauty of biotic communities is what is most fundamentally to be conserved. In a comprehensive ethics of respect for life, we ought to set ethics at the level of ecosystems alongside classical, humanistic ethics.

Value Theory

In practice the ultimate challenge of environmental ethics

is the conservation of life on Earth. In principle the ultimate challenge is a value theory profound enough to support that ethic. In nature there is negentropic construction in dialectic with entropic teardown, a process for which we hardly yet have an adequate scientific, much less a valuational theory. Yet this is nature's most striking feature, one that ultimately must be valued and of value. In one sense nature is indifferent to mountains, rivers, fauna, flora, forests and grasslands. But in another sense nature has bent toward making and remaking these projects, millions of kinds, for several billion years.

These performances are worth noticing--remarkable, memorable--and not just because of their tendencies to produce something else, certainly not merely because of their tendency to produce this noticing in certain recent subjects--our human selves. These events are loci of value as products of systemic nature in its formative processes. The splendors of Earth do not simply lie in their roles as human resources, supports of culture, or stimulators of experience. The most plausible account will find some programmatic evolution toward value, and not because it ignores Darwin but because it heeds his principle of natural selection and deploys this into a selection exploring new niches and elaborating kinds, even a selection upslope toward higher values, at least along some trends within some ecosystems.

A systematic environmental ethics does not wish to believe in the special creation of values, nor in their dumfounding epigenesis. The notion that nature is a value carrier is ambiguous. Much depends on a thing being more or less structurally congenial for the carriage. We value a thing to discover that we are under the sway of its valence, inducing our behavior. It has among its "strengths" this capacity to carry value. This potential cannot always be of the empty sort that a glass has for carrying water. It is often pregnant fullness. Some of the values that nature carries are up to us, our assignment. But fundamentally there are powers in nature that move to us and through us. There is no value without an evaluator. So runs a well-entrenched dogma.

Humans clearly evaluate their world; sentient animals may also. But plants cannot evaluate their environment; they have no options and make no choices. A fortiori, species and ecosystems, Earth and Nature cannot be bona fide evaluators. One can always hang on to the claim that value, like a tickle or remorse, must be felt to be there. Its esse is percipi. Nonsensed value is nonsense. There are no thoughts without a thinker, no percepts without a perceiver, no deeds without a doer, no targets without an aimer. Such resolute subjectivists cannot be defeated by argument, although they can be driven toward analyticity. That theirs is a retreat to definition is difficult to expose, because they seem to cling so closely to inner experience. They are reporting, on this hand, how values always excite us. They are giving, on that hand, a stipulative definition. That is how they choose to use the word "value."

If value arrives only with consciousness, experiences where humans find value there have to be dealt with as appearances of various sorts. The value has to be relocated in the valuing subject's creativity as a person meets a valueless world, or even a valuable one--one able to be valued--but which before the human bringing of value ability contains only possibility and not any actual value. Value can only be extrinsic to nature, never intrinsic to it. But the valuing subject in an otherwise valueless world is an insufficient premise for the experienced conclusions of those who respect all life.

Conversion to a biological view seems truer to world experience and more logically compelling. Here the order of knowing reverses--and also enhances--the order of being. This too is a perspective, but ecologically better informed. Science has been steadily showing how the consequents (life, mind) are built on their precedents (energy, matter), however much they overleap them. Life and mind appear where they did not before exist, and with this levels of value emerge that did not before exist. But that gives no reason to say that all value is an irreducible emergent at the human (or upper animal) level.

A comprehensive environmental ethics reallocates value across the whole continuum. Value increases in the emergent climax, but is continuously present in the composing precedents. The system is value-able, able to produce value. Human evaluators are among its products. Some value depends on subjectivity, yet all value is generated within the geosystemic and ecosystemic pyramid. Systemically, value fades from subjective to objective value, but also fans out from the individual to its role and matrix. Things do not have their separate natures merely in and for themselves, but they face outward and co-fit into broader natures.

Value-in-itself is smeared out to become value-in-togetherness. Value seeps out into the system, and we lose our capacity to identify the individual as the sole locus of value. Intrinsic value, that of an individual "for what it is in itself," becomes problematic in a holistic web. True, the system produces such values more and more with its evolution of individuality and freedom. Yet to decouple this from the biotic, communal system is to make value too internal and elementary; this forgets relatedness and externality. Every intrinsic value has leading and trailing and's pointing to value from which it comes and toward which it moves.

Adapted fitness makes individualistic value too system independent. Intrinsic value is a part in a whole, not to be fragmented by valuing it in isolation. Everything is good in a role, in a whole, although we can speak of objective intrinsic goodness wherever a point event--a trillium--defends a good (its life) in itself. We can speak of subjective intrinsic goodness when such an event registers as a point experience, at which point humans pronounce both their experience and what it is of good without need to enlarge their focus. Neither the trilliums nor the human judges of it require for their respective valuings any further contributory reference.

When eaten by foragers or in death resorbed into humus, the trillium has its value destroyed, transformed into instrumentality. The system is a value transformer where form and being, process and reality, fact and value are inseparably joined. Intrinsic and instrumental values shuttle back and forth, parts-in-wholes and wholes-in-parts, local details of value embedded in global structures, gems in their settings, and their setting-situation a corporation where value cannot stand alone. Every good is in community.

In environmental ethics one's beliefs about nature, which are based upon but exceed science, have everything to do with beliefs about duty. The way the world is informs the way it ought to be. We always shape our values in significant measure in accord with our notion of the kind of universe that we live in, and this drives our sense of duty. Our model of reality implies a model of conduct. Differing models sometimes imply similar conduct, but often they do not.

A model in which nature has no value apart from human preferences will imply different conduct from one where nature projects fundamental values, some objective and others that further require human subjectivity superposed on objective nature. This evaluation is not scientific description; hence not ecology per se, but metaecology. No amount of research can verify that, environmentally, the right is the optimum biotic community. Yet ecological description generates this valuing of nature, endorsing the systemic rightness.

The transition from is to good and hence to ought occurs here; we leave science to enter the domain of evaluation, from which an ethic follows. What is ethically puzzling and exciting is that an ought is not so much derived from an is as discovered simultaneously with it. As we progress from descriptions of fauna and flora, of cycles and pyramids, of autotrophs coordinated with heterotrophs, of stability and dynamism, on to intricacy, planetary opulence and interdependence, to unity and harmony with oppositions in counterpoint and synthesis, organisms evolved within and satisfactorily fitting their communities, arriving at length at beauty and goodness, it is difficult to say where the natural facts leave off and where the natural values appear.

For some at least, the sharp is/ought dichotomy is gone; the values seem to be there as soon as the facts are fully in, and both alike properties of the system. There is something overspecialized about an ethic, held by the dominant class of Homo sapiens, that regards the welfare of only one of several million species as an object and beneficiary of duty. If this requires a paradigm change about the sorts of things to which duty can attach, so much the worse for those humanistic ethics no longer functioning in, nor suited to, their changing environment.

The anthropocentrism associated with them was fiction anyway. There is something Newtonian, not yet Einsteinian, besides something morally naive, about living in a reference frame where one species takes itself as absolute and values everything else relative to its utility. Only the human species contains moral agents, but perhaps conscience on such an Earth ought not be used to exempt every other form of life from consideration, with the resulting paradox that the sole moral species acts only in its collective self-interest toward all the rest.

9

The Environment (Protection) Act, 1986

An Act to provide for the protection and improvement of environment and for matters connected there with: Whereas the decisions were taken at the United Nations Conference on the Human Environment held at Stockholm in June, 1972, in which India participated, to take appropriate steps for the protection and improvement of human environment;

And whereas it is considered necessary further to implement the decisions aforesaid in so far as they relate to the protection and improvement of environment and the prevention of hazards to human beings, other living creatures, plants and property; BE it enacted by Parliament in the Thirty-seventh Year of the Republic of India as follows:-

Preliminary

Short Title, Extend and Commencement

1. This Act may be called the Environment (Protection) Act, 1986.
2. It extends to the whole of India.

3. It shall come into force on such date as the Central Government may, by notification in the Official Gazette, appoint and different dates may be appointed for different provisions of this Act and for different areas.

Definitions

In this Act, unless the context otherwise requires:-

(a) "environment" includes water, air and land and the inter-relationship which exists among and between water, air and land, and human beings, other living creatures, plants, micro-organism and property;

(b) "environmental pollutant" means any solid, liquid or gaseous substance present in such concentration as may be, or tend to be, injurious to environment;

(c) "environmental pollution" means the presence in the environment of any environmental pollutant;

(d) "handling", in relation to any substance, means the manufacture, processing, treatment, package, storage, transportation, use, collection, destruction, conversion, offering for sale, transfer or the like of such substance;

(e) "hazardous substance" means any substance or preparation which, by reason of its chemical or physico-chemical properties or handling, is liable to cause harm to human beings, other living creatures, plant, micro-organism, property or the environment;

(f) "occupier", in relation to any factory or premises, means a person who has, control over the affairs of the factory or the premises and includes in relation to any substance, the person in possession of the substance;

(g) "prescribed" means prescribed by rules made under this Act.

General Powers of Central Government

Power of Central Government to Take Measures to Protect and Improve Environment

Subject to the provisions of this Act, the Central Government, shall have the power to take all such measures as it deems necessary or expedient for the purpose of protecting and improving the quality of the environment and preventing controlling and abating environmental pollution.

In particular, and without prejudice to the generality of the provisions of sub-section (1), such measures may include measures with respect to all or any of the following matters, namely:-

(i) co-ordination of actions by the State Governments, officers and other authorities--

— under this Act, or the rules made thereunder, or

— under any other law for the time being in force which is relatable to the objects of this Act;

(ii) planning and execution of a nation-wide programme for the prevention, control and abatement of environmental pollution;

(iii) laying down standards for the quality of environment in its various aspects;

(iv) laying down standards for emission or discharge of environmental pollutants from various sources whatsoever:

Provided that different standards for emission or discharge may be laid down under this clause from different sources having regard to the quality or composition of the emission or discharge of environmental pollutants from such sources;

(v) restriction of areas in which any industries, operations or processes or class of industries, operations or processes shall not be carried out or shall be carried out subject to certain safeguards;

(vi) laying down procedures and safeguards for the prevention of accidents which may cause environmental pollution and remedial measures for such accidents;

(vii) laying down procedures and safeguards for the handling of hazardous substances;

(viii) examination of such manufacturing processes, materials and substances as are likely to cause environmental pollution;

(ix) carrying out and sponsoring investigations and research relating to problems of environmental pollution;

(x) inspection of any premises, plant, equipment, machinery, manufacturing or other processes, materials or substances and giving, by order, of such directions to such authorities, officers or persons as it may consider necessary to take steps for the prevention, control and abatement of environmental pollution;

(xi) establishment or recognition of environmental laboratories and institutes to carry out the functions entrusted to such environmental laboratories and institutes under this Act;

(xii) collection and dissemination of information in respect of matters relating to environmental pollution;

(xiii) preparation of manuals, codes or guides relating to the prevention, control and abatement of environmental pollution;

(xiv)such other matters as the Central Government deems necessary or expedient for the purpose of securing the effective implementation of the provisions of this Act.

The Central Government may, if it considers it necessary or expedient so to do for the purpose of this Act, by order, published in the Official Gazette, constitute an authority or authorities by such name or names as may be specified in the order for the purpose of exercising and performing such of the powers and functions (including the power to issue directions under section 5) of the Central Government under this Act and for taking measures with respect to such of the matters referred to in sub-section (2) as may be mentioned in the order and subject to the supervision and control of the Central Government and the provisions of such order, such authority or authorities may exercise and powers or perform the functions or take the measures so mentioned in the order as if such authority or authorities had been empowered by this Act to exercise those powers or perform those functions or take such measures.

Appointment of Officers and their Powers and Functions

— Without prejudice to the provisions of sub-section (3) of section 3, the Central Government may appoint officers with such designation as it thinks fit for the purposes of this Act and may entrust to them such of the powers and functions under this Act as it may deem fit.

— The officers appointed under sub-section (1) shall be subject to the general control and direction of the Central Government or, if so directed by that Government, also of the authority or authorities, if any, constituted under sub- section (3) of section 3 or of any other authority or officer.

Power to Give Directions

Notwithstanding anything contained in any other law but subject to the provisions of this Act, the Central Government may, in the exercise of its powers and performance of its functions under this Act, issue directions in writing to any person, officer or any authority and such person, officer or authority shall be bound to comply with such directions.

Explanation: For the avoidance of doubts, it is hereby declared that the power to issue directions under this section includes the power to direct:-

(a) the closure, prohibition or regulation of any industry, operation or process; or

(b) stoppage or regulation of the supply of electricity or water or any other service.

Rules to Regulate Environmental Pollution

(1) The Central Government may, by notification in the Official Gazette, make rules in respect of all or any of the matters referred to in section 3.

(2) In particular, and without prejudice to the generality of the foregoing power, such rules may provide for all or any of the following matters, namely:-

  (a) the standards of quality of air, water or soil for various areas and purposes;

  (b) the maximum allowable limits of concentration of various environmental pollutants (including noise) for different areas;

  (c) the procedures and safeguards for the handling of hazardous substances;

  (d) the prohibition and restrictions on the handling of hazardous substances in different areas;

(e) the prohibition and restriction on the location of industries and the carrying on process and operations in different areas;

(f) the procedures and safeguards for the prevention of accidents which may cause environmental pollution and for providing for remedial measures for such accidents.

Prevention, Control, and Abatement of Environmental Pollution

Persons carrying on industry operation, etc., not to allow emission or discharge of environmental pollutants in excess of the standards:

No person carrying on any industry, operation or process shall discharge or emit or permit to be discharged or emitted any environmental pollutants in excess of such standards as may be prescribed.

Persons handling hazardous substances to comply with procedural safeguards:

No person shall handle or cause to be handled any hazardous substance except in accordance with such procedure and after complying with such safeguards as may be prescribed.

Furnishing of information to authorities and agencies in certain cases:

— Where the discharge of any environmental pollutant in excess of the prescribed standards occurs or is apprehended to occur due to any accident or other unforeseen act or event, the person responsible for such discharge and the person in charge of the place at which such discharge occurs or is apprehended to occur shall be bound to prevent or mitigate the environmental pollution caused as a result of such discharge and shall also forthwith:-

(a) intimate the fact of such occurrence or apprehension of such occurrence; and

(b) be bound, if called upon, to render all assistance, to such authorities or agencies as may be prescribed.

— On receipt of information with respect to the fact or apprehension on any occurrence of the nature referred to in sub-section (1), whether through intimation under that sub-section or otherwise, the authorities or agencies referred to in sub-section (1) shall, as early as practicable, cause such remedial measures to be taken as necessary to prevent or mitigate the environmental pollution.

— The expenses, if any, incurred by any authority or agency with respect to the remedial measures referred to in sub-section (2), together with interest (at such reasonable rate as the Government may, by order, fix) from the date when a demand for the expenses is made until it is paid, may be recovered by such authority or agency from the person concerned as arrears of land revenue or of public demand.

Powers of Entry and Inspection

— Subject to the provisions of this section, any person empowered by the Central Government in this ehalf shall have a right to enter, at all reasonable times with such assistance as he considers necessary, any place-

(a) for the purpose of performing any of the functions of the Central Government entrusted to him;

(b) for the purpose of determining whether and if so in what manner, any such functions are to be performed or whether any provisions of this Act or the rules made thereunder orany notice, order, direction or authorisation served, made, given or granted under this Act is being or has been complied with;

(c) for the purpose of examining and testing any equipment, industrial plant, record, register, document or any other material object or for conducting a search of any building in which he has reason to believe that an offence under this Act or the rules made thereunder has been or is being or is about to be committed and for seizing any such equipment, industrial plant, record, register, document or other material object if he has reason to believe that it may furnish evidence of the commission of an offence punishable under this Act or the rules made thereunder or that such seizure is necessary to prevent or mitigate environmental pollution.

— Every person carrying on any industry, operation or process of handling any hazardous substance shall be bound to render all assistance to the person empowered by the Central Government under sub-section (1) for carrying out the functions under that sub-section and if he fails to do so without any reasonable cause or excuse, he shall be guilty of an offence under this Act.

— If any person wilfully delays or obstructs any persons empowered by the Central Government under sub-section (1) in the performance of his functions, he shall be guilty of an offence under this Act.

— The provisions of the Code of Criminal Procedure, 1973, or, in relation to the State of Jammu and Kashmir, or an area in which that Code is not in force, the provisions of any corresponding law in force in that State or area shall, so far as may be, apply to any search or seizures under this section as they apply to any search or seizure made under the authority of a warrant issued under section 94 of the said Code or as the case may be, under the corresponding provision of the said law.

Power to Take Sample and Procedure to be Followed in Connection Therewith

— The Central Government or any officer empowered by it in this behalf, shall have power to take, for the purpose of analysis, samples of air, water, soil or other substance from any factory, premises or other place in such manner as may be prescribed.

— The result of any analysis of a sample taken under sub-section (1) shall not be admissible in evidence in any legal proceeding unless the provisions of sub-sections (3) and (4) are complied with.

— Subject to the provisions of sub-section (4), the person taking the sample under sub-section (1) shall:-

(a) serve on the occupier or his agent or person in charge of the place, a notice, then and there, in such form as may be prescribed, of his intention to have it so analysed;

(b) in the presence of the occupier of his agent or person, collect a sample for analysis;

(c) cause the sample to be placed in a container or containers which shall be marked and sealed and shall also be signed both by the person taking the sample and the occupier or his agent or person;

(d) send without delay, the container or the containers to the laboratory established or recognised by the Central Government under section 12.

— When a sample is taken for analysis under sub-section (1) and the person taking the sample serves on the occupier or his agent or person, a notice under clause (a) of sub-section (3), then,:-

(a) in a case where the occupier, his agent or person wilfully absents himself, the person taking the sample shall collect the sample for analysis to be placed in a container or containers which shall be marked and sealed and shall also be signed by the person taking the sample, and

(b) in a case where the occupier or his agent or person present at the time of taking the sample refuses to sign the marked and sealed container or containers of the sample as required under clause

(c) of sub-section (3), the marked and sealed container or containers shall be signed by the person taking the samples, and the container or containers shall be sent without delay by the person taking the sample for analysis to the laboratory established or recognised under section 12 and such person shall inform the Government Analyst appointed or recognised under section 12 in writing, about the wilfull absence of the occupier or his agent or person, or, as the case may be, his refusal to sign the container or containers.

Environmental Laboratories

— The Central Government may, by notification in the Official Gazette,:-

(a) establish one or more environmental laboratories;

(b) recognise one or more laboratories or institutes as environmental laboratories to carry out the functions entrusted to an environmental laboratory under this Act.

— The Central Government may, by notification in the Official Gazette, make rules specifying:

(a) the functions of the environmental laboratory;

(b) the procedure for the submission to the said laboratory of samples of air, water, soil or other substance for analysis .or tests, the form of the laboratory report thereon and the fees payable for such report;

(c) such other matters as may be necessary or expedient to enable that laboratory to carry out its functions.

Government Analysts

The Central Government may by notification in the Official Gazette, appoint or recognise such persons as it thinks fit and having the prescribed qualifications to be Government Analysts for the purpose of analysis of samples of air, water, soil or other substance sent for analysis to any environmental laboratory established or recognised under sub-section (1) of section 12.

Reports of Government Analysts

Any document purporting to be a report signed by a Government analyst may be used as evidence of the facts stated therein in any proceeding under this Act.

Penalty for contravention of the provisions of the act and the rules, orders and directions:

— Whoever fails to comply with or contravenes any of the provisions of this Act, or the rules made or orders or directions issued thereunder, shall, in respect of each such failure or contravention, be punishable with imprisonment for a term which may extend to five years with fine which may extend to one lakh rupees, or with both, and in case the failure or contravention continues, with additional fine which may extend to five thousand rupees for every day during which such failure or contravention continues after the conviction for the first such failure or contravention.

— If the failure or contravention referred to in sub-section (1) continues beyond a period of one year after the date of conviction, the offender shall be punishable with imprisonment for a term which may extend to seven years.

Offences by Companies

— Where any offence under this Act has been committed by a company, every person who, at the time the offence was committed, was directly in charge of, and was responsible to, the company for the conduct of the business of the company, as well as the company, shall be deemed to be guilty of the offence and shall be liable to be proceeded against and punished accordingly:

Provided that nothing contained in this sub-section shall render any such person liable to any punishment provided in this Act, if he proves that the offence was committed without his knowledge or that he exercised all due diligence to prevent the commission of such offence.

— Notwithstanding anything contained in sub-section (1), where an offence under this Act has been committed by a company and it is proved that the offence has been committed with the consent or connivance of, or is attributable to any neglect on the part of, any director, manager, secretary or other officer of the company, such director, manager, secretary or other officer shall also deemed to be guilty of that offence and shall be liable to be proceeded against and punished accordingly. Explanation--For the purpose of this section,:-

(a) "company" means any body corporate and includes a firm or other association of individuals;

(b) "director", in relation to a firm, means a partner in the firm.

Offences by Government Departments

— Where an offence under this Act has been committed by any Department of Government, the Head of the Department shall be deemed to be guilty of the offence and shall be liable to be proceeded against and punished accordingly.

Provided that nothing contained in this section shall render such Head of the Department liable to any punishment if he proves that the offence was committed without his knowledge or that he exercise all due diligence to prevent the commission of such offence.

— Notwithstanding anything contained in sub-section (1), where an offence under this Act has been committed by a Department of Government and it is proved that the offence has been committed with the consent or connivance of, or is attributable to any neglect on the part of, any officer, other than the Head of the Department, such officer shall also be deemed to be guilty of that offence and shall be liable to be proceeded against and punished accordingly.

MISCELLANEOUS

Protection of action taken in good faith

No suit, prosecution or other legal proceeding shall lie against the Government or any officer or other employee of the Government or any authority constituted under this Act or any member, officer or other employee of such authority in respect of anything which is done or intended to be done in good faith in pursuance of this Act or the rules made or orders or directions issued thereunder.

Cognizance of Offences

No court shall take cognizance of any offence under this Act except on a complaint made by:-

— the Central Government or any authority or officer authorised in this behalf by that Government, or

— any person who has given notice of not less than sixty days, in the manner prescribed, of the alleged offence and of his intention to make a complaint, to the Central Government or the authority or officer authorised as aforesaid.

Information, Reports or Returns

The Central Government may, in relation to its function under this Act, from time to time, require any person, officer, State Government or other authority to furnish to it or any prescribed authority or officer any reports, returns, statistics, accounts and other information and such person, officer, State Government or other authority shall be bound to do so.

Members, Officers and Employees of the Authority

Constituted under section 3 to be public servants: All the members of the authority, constituted, if any, under section 3 and all officers and other employees of such authority when acting or purporting to act in pursuance of any provisions of this Act or the rules made or orders or directions issued thereunder shall be deemed to be public servants within the meaning of section 21 of the Indian Penal Code (45 of 1860).

Bar of Jurisdiction

No civil court shall have jurisdiction to entertain any suit or proceeding in respect of anything done, action taken or order or direction issued by the Central Government or any other authority or officer in pursuance of any power conferred by or in relation to its or his functions under this Act.

Powers to Delegate

Without prejudice to the provisions of sub-section (3) of section 3, the Central Government may, by notification in the Official Gazette, delegate, subject to such conditions and limitations as may be specified in the notifications, such of its powers and functions under this Act [except the powers to constitute an authority under sub-section (3) of section 3 and to make rules under section 25] as it may deem necessary or expedient, to any officer, State Government or other authority.

Effect of Other Laws

— Subject to the provisions of sub-section (2), the provisions of this Act and the rules or orders made therein shall have effect notwithstanding anything inconsistent therewith contained in any enactment other than this Act.

— Where any act or omission constitutes an offence punishable under this Act and also under any other Act then the offender found guilty of such offence shall be liable to be punished under the other Act and not under this Act.

Power to Make Rules

— The Central Government may, by notification in the

Official Gazette, make rules for carrying out the purposes of this Act.

— In particular, and without prejudice to the generality of the foregoing power, such rules may provide for all or any of the following matters, namely:-

(a) the standards in excess of which environmental pollutants shall not be discharged or emitted under section 7;

(b) the procedure in accordance with and the safeguards in compliance with which hazardous substances shall be handled or caused to be handled under section 8;

(c) the authorities or agencies to which intimation of the fact of occurrence or apprehension of occurrence of the discharge of any environmental pollutant in excess of the prescribed standards shall be given and to whom all assistance shall be bound to be rendered under sub-section (1) of section 9;

(d) the manner in which samples of air, water, soil or other substance for the purpose of analysis shall be taken under sub-section (1) of section 11;

(e) the form in which notice of intention to have a sample analysed shall be served under clause (a) of sub section (3) of section 11;

(f) the functions of the environmental laboratories, the procedure for the submission to such laboratories of samples of air, water, soil and other substances for analysis or test; the form of laboratory report; the fees payable for such report and other matters to enable such laboratories to carry out their functions under sub-section (2) of section 12;

(g) the qualifications of Government Analyst appointed or recognised for the purpose of analysis of samples of air, water, soil or other substances under section 13;

(h) the manner in which notice of the offence and of the intention to make a complaint to the Central Government shall be given under clause (b) of section 19;

(i) the authority of officer to whom any reports, returns, statistics, accounts and other information shall be furnished under section 20;

(j) any other matter which is required to be, or may be,prescribed.

Rules Made Under this Act to be Laid before Parliament

Every rule made under this Act shall be laid, as soon as may be after it is made, before each Hose of Parliament, while it is in session, for a total period of thirty days which may be comprised in one session or in two or more successive sessions, and if, before the expiry of the session immediately following the session or the successive sessions aforesaid, both Houses agree in making any modification in the rule or both Houses agree that the rule should not be made, the rule shall thereafter have effect only in such modified form or be of no effect, as the case may be; so, however, that any such modification or annulment shall be without prejudice to the validity of anything previously done under that rule.

10

The Earth Summit

The Earth Summit in Rio de Janeiro was unprecedented for a UN conference, in terms of both its size and the scope of its concerns. Twenty years after the first global environment conference, the UN sought to help Governments rethink economic development and find ways to halt the destruction of irreplaceable natural resources and pollution of the planet. Hundreds of thousands of people from all walks of life were drawn into the Rio process. They persuaded their leaders to go to Rio and join other nations in making the difficult decisions needed to ensure a healthy planet for generations to come.

The Summit's message - that nothing less than a transformation of our attitudes and behaviour would bring about the necessary changes - was transmitted by almost 10,000 on-site journalists and heard by millions around the world. The message reflected the complexity of the problems facing us: that poverty as well as excessive consumption by affluent populations place damaging stress on the environment. Governments recognized the need to redirect international and national plans and policies to ensure that all economic decisions fully took into account any environmental impact. And the message has produced results, making eco-efficiency a guiding principle for business and governments alike.

— Patterns of production: particularly the production of toxic components, such as lead in gasoline, or poisonous waste - are being scrutinized in a systematic manner by the UN and Governments alike;

— Alternative sources of energy are being sought to replace the use of fossil fuels which are linked to global climate change;

— New reliance on public transportation systems is being emphasized in order to reduce vehicle emissions, congestion in cities and the health problems caused by polluted air and smog;

— There is much greater awareness of and concern over the growing scarcity of water.

The two-week Earth Summit was the climax of a process, begun in December 1989, of planning, education and negotiations among all Member States of the United Nations, leading to the adoption of Agenda 21, a wide-ranging blueprint for action to achieve sustainable development worldwide. At its close, Maurice Strong, the Conference Secretary-General, called the Summit a "historic moment for humanity".

Although Agenda 21 had been weakened by compromise and negotiation, he said, it was still the most comprehensive and, if implemented, effective programme of action ever sanctioned by the international community. Today, efforts to ensure its proper implementation continue, and they will be reviewed by the UN General Assembly at a special session to be held in June 1997.

The Earth Summit influenced all subsequent UN conferences, which have examined the relationship between human rights, population, social development, women and human settlements - and the need for environmentally sustainable development. The World Conference on Human Rights, held in Vienna in 1993, for example, underscored the right of people to a healthy environment and the right to development, controversial demands that had met with resistance from some Member States until Rio.

The relationship between economic development and environmental degradation was first placed on the international agenda in 1972, at the UN Conference on the Human Environment, held in Stockholm. After the Conference, Governments set up the United Nations Environment Programme (UNEP), which today continues to act as a global catalyst for action to protect the environment. Little, however, was done in the succeeding years to integrate environmental concerns into national economic planning and decision-making. Overall, the environment continued to deteriorate, and such problems as ozone depletion, global warming and water pollution grew more serious, while the destruction of natural resources accelerated at an alarming rate.

By 1983, when the UN set up the World Commission on Environment and Development, environmental degradation, which had been seen as a side effect of industrial wealth with only a limited impact, was understood to be a matter of survival for developing nations. Led by Gro Harlem Brundtland of Norway, the Commission put forward the concept of sustainable development as an alternative approach to one simply based on economic growth - one "which meets the needs of the present without compromising the ability of future generations to meet their own needs".

After considering the 1987 Brundtland report, the UN General Assembly called for the UN Conference on Environment and Development (UNCED). The primary goals of the Summit were to come to an understanding of "development" that would support socio-economic development and prevent the continued deterioration of the environment, and to lay a foundation for a global partnership between the developing and the more industrialized countries, based on mutual needs and common interests, that would ensure a healthy future for the planet.

Earth Summit Agreements

In Rio, Governments - 108 represented by heads of State or Government - adopted three major agreements aimed at changing the traditional approach to development:

— Agenda 21 - a comprehensive programme of action for global action in all areas of sustainable development;

— The Rio Declaration on Environment and Development - a series of principles defining the rights and responsibilities of States;

— The Statement of Forest Principles - a set of principles to underlie the sustainable management of forests worldwide.

In addition, two legally binding Conventions aimed at preventing global climate change and the eradication of the diversity of biological species were opened for signature at the Summit, giving high profile to these efforts:

— The United Nations Framework Convention on Climate Change and

— The Convention on Biological Diversity

Agenda 21 addresses today's pressing problems and aims to prepare the world for the challenges of the next century. It contains detailed proposals for action in social and economic areas (such as combating poverty, changing patterns of production and consumption and addressing demographic dynamics), and for conserving and managing the natural resources that are the basis for life - protecting the atmosphere, oceans and biodiversity; preventing deforestation; and promoting sustainable agriculture, for example.

Governments agreed that the integration of environment and development concerns will lead to the fulfilment of basic needs, improved standards for all, better protected and better managed ecosystems and a safer and a more prosperous future. "No nation can achieve this on its own. Together we can - in a global partnership for sustainable development", states the preamble.

The programme of action also recommends ways to strengthen the part played by major groups - women, trade unions, farmers, children and young people, indigenous peoples, the scientific community, local authorities, business, industry and non-Governmental Organizations (NGOs) - in achieving sustainable development.

The Rio Declaration on Environment and Developmentsupports Agenda 21 by defining the rights and responsibilities of States regarding these issues. Among its principles:

— That human beings are at the centre of concerns for sustainable development. They are entitled to a healthy and productive life in harmony with nature;

— That scientific uncertainty should not delay measures to prevent environmental degradation where there are threats of serious or irreversible damage;

— That States have a sovereign right to exploit their own resources but not to cause damage to the environment of other States;

— That eradicating poverty and reducing disparities in worldwide standards of living are "indispensable" for sustainable development;

— That the full participation of women is essential for achieving sustainable development; and

— That the developed countries acknowledge the responsibility that they bear in the international pursuit of sustainable development in view of the pressures their societies place on the global environment and of the technologies and financial resources they command.

The Statement of Forest Principles, the non-legally binding statement of principles for the sustainable management of forests, was the first global consensus reached on forests. Among its provisions:

— That all countries, notably developed countries, should make an effort to "green the world" through reforestation and forest conservation;

— That States have a right to develop forests according to their socio-economic needs, in keeping with national sustainable development policies; and

— That specific financial resources should be provided to develop programmes that encourage economic and social substitution policies.

At the Summit, the UN was also called on to negotiate an international legal agreement on desertification, to hold talks on preventing the depletion of certain fish stocks, to devise a programme of action for the sustainable development of small island developing States and to establish mechanisms for ensuring the implementation of the Rio accords.

UN Follow-Up

The Earth Summit succeeded in presenting new perspectives on economic progress. It was lauded as the beginning of a new era and its success would be measured by the implementation - locally, nationally and internationally - of its agreements. Those attending the Summit understood that making the necessary changes would not be easy: it would be a multi-phased process; it would take place at different rates in different parts of the world; and it would require the expenditure of funds now in order to prevent much larger financial and environmental costs in the future.

In Rio, the UN was given a key role in the implementation of Agenda 21. Since then, the Organization has taken steps to integrate concepts of sustainable development into all relevant policies and programmes. Income-generating projects increasingly take into account environmental consequences. Development assistance programmes are increasingly directed towards women, given their central roles as producers and as caretakers of families.

Efforts to manage forests in a sustainable manner begin with finding alternatives to meet the needs of people who are overusing them. The moral and social imperatives for alleviating poverty are given additional urgency by the recognition that poor people can cause damage to the environment. And foreign investment decisions increasingly take into account the fact that drawing down the earth's natural resources for short-term profit is bad for business in the long run.

In adopting Agenda 21, the Earth Summit also requested the United Nations to initiate talks aimed at halting the rapid depletion of certain fish stocks and preventing conflict over fishing on the high seas. After negotiations spanning more than two years, the UN Agreement on High Seas Fishing was opened for signature on 4 December 1995. It provides for all species of straddling and highly migratory fish - those which swim between national economic zones or migrate across broad areas of the ocean - to be subject to quotas designed to ensure the continued survival of fish for our children and grandchildren to enjoy.

Also at the Summit, Governments requested the UN to hold negotiations for an international legal agreement to prevent the degradation of drylands. The resulting International Convention to Combat Desertification in Those Countries Experiencing Serious Drought and/or Desertification, particularly in Africa, was opened for signing in October 1994 and entered into force in December 1996. It calls for urgent action to be taken in Africa, where some 66 per cent of the continent is desert or drylands and 73 per cent of agricultural drylands are already degraded.

In order to promote the well-being of people living in island countries, the Summit called for the UN to convene a Global Conference on the Sustainable Development of Small Island Developing States . The Conference was held in Barbados in May 1994 and produced a programme of action designed to assist these environmentally and economically vulnerable countries.

In addition, three bodies were created within the United Nations to ensure full support for implementation of Agenda 21 worldwide:

— The UN Commission on Sustainable Development, which first met in June 1993;

— The Inter-agency Committee on Sustainable Development, set up by the Secretary-General in 1992 to ensure effective system-wide cooperation and coordination in the follow-up to the Summit; and

— The High-level Advisory Board on Sustainable Development, established in 1993 to advise the Secretary-General and the Commission on issues relating to the implementation of Agenda 21.

UN Commission on Sustainable Development (CSD) - The Earth Summit called on the General Assembly to establish the Commission under the Economic and Social Council as a means of supporting and encouraging action by Governments, business, industry and other non-governmental groups to bring about the social and economic changes needed for sustainable development. Each year, the Commission reviews implementation of the Earth Summit agreements, provides policy guidance to Governments and major groups involved in sustainable development and strengthens Agenda 21 by devising additional strategies where necessary.

It also promotes dialogue and builds partnerships between Governments and the major groups which are seen as key to achieving sustainable development worldwide. The work of the Commission was supported by numerous inter-sessional meetings and activities initiated by Governments, international organizations and major groups. In June 1997, the General Assembly will hold a special session to review overall progress following the Earth Summit.

Under a multi-year thematic work programme, the Commission has monitored the early implementation of Agenda 21 in stages. Each sectoral issue - health, human settlements, freshwater, toxic chemicals and hazardous waste, land, agriculture, desertification, mountains, forests, biodiversity, atmosphere, oceans and seas - was reviewed between 1994 and 1996. Developments on most "cross-sectoral" issues are considered each year.

These issues, which must be addressed if action in sectoral areas is to be effective, are clustered as follows: critical elements of sustainability (trade and environment, patterns of production and consumption, combating poverty, demographic dynamics); financial resources and mechanisms; education, science, transfer of environmentally sound technologies, technical cooperation and capacity-building; decision-making; and activities of the major groups, such as business and labour.

In 1995, the Commission established under its auspices the Intergovernmental Panel on Forests with a broad mandate covering the entire spectrum of forest-related issues and dealing with conservation, sustainable development and management of all types of forests. The Panel will submit its final report containing concrete conclusions and proposals for action to the 1997 session of the CSD.

Reports submitted annually by Governments are the main basis for monitoring progress and identifying problems faced by countries. By mid-1996, some 100 Governments had established national sustainable development councils or other coordinating bodies. More than 2,000 municipal and town governments had each formulated a local Agenda 21 of its own. Many countries were seeking legislative approval for sustainable development plans, and the level of NGO involvement remained high.

Standard-setting

Central to the ability of Governments to formulate policies for sustainability and to regulate their impact is the development of a set of internationally accepted criteria and indicators for sustainable development. The Commission on Sustainable Development is spearheading the work, which will enable countries to gather and report the data needed to measure progress on Agenda 21. It is hoped that a "menu" of indicators - from which Governments will choose those appropriate to local conditions - will be used by countries in their national plans and strategies and, subsequently, when they report to the Commission.

Achieving sustainable development worldwide depends largely on changing patterns of production and consumption - what we produce, how it is produced and how much we consume, particularly in the developed countries. CSD's work programme in this area focuses on projected trends in consumption and production; impacts on developing countries, including trade opportunities; assessment of the effectiveness of policy instruments, including new and innovative instruments; progress by countries through their timebound voluntary commitments; and extension and revision of UN guidelines for consumer protection.

In 1995, the Commission also adopted a work programme on the transfer of environmentally sound technology, cooperation and capacity building. The programme places an emphasis on three interrelated priority areas: access to and dissemination of information, capacity building for managing technological change, and financial and partnership arrangements. The Commission is working with the World Trade Organization, the UN Conference on Trade and Development and the United Nations Environment Programme (UNEP) to ensure that trade, environment and sustainable development issues are mutually reinforcing.

Financing Sustainable Development

At Rio, it was agreed that most financing for Agenda 21 would come from within a country's own public and private sectors. However, new and additional external funds were considered necessary if developing countries were to adopt sustainable development practices. Of the estimated $600 billion required annually by developing countries to implement Agenda 21, most - $475 billion - was to be transferred from economic activities in those countries.

A further $125 billion would be needed in new and additional funds from external sources, some $70 billion more than current levels of official development assistance

(ODA). According to the Organisation for Economic Co-operation and Development (OECD), between 1992 and 1995, levels of ODA fell from about $60.8 billion to $59.2 billion, despite a call at Rio for donor countries to more than double their official assistance.

Other monies are available for implementation of Agenda 21. The Global Environment Facility (GEF) was set up in 1991. It is implemented by the World Bank, the United Nations Development Programme and the United Nations Environment Programme. The GEF provides funding for activities aimed at achieving global environmental benefits in four areas: climate change, loss of biodiversity, pollution of international waters and the depletion of the ozone layer.

At Rio, the Facility became the funding mechanism for activities under the United Nations Framework Convention on Climate Change and the Convention on Biological Diversity. In 1994, the scope of the GEF's funding was broadened to include land degradation, primarily desertification and deforestation, where this is linked to the four focal areas above. Since 1992, some $2 billion has been pledged for activities supported by the GEF.

In the years since the Earth Summit, the level of funding channelled to many of the developing countries as direct private investment has increased significantly and now far outstrips official flows. In 1995, this reportedly amounted to some $95 billion. Efforts are being made to ensure that activities supported by these funds are also environmentally sustainable.

Five Years After Rio

In June 1997, the world's attention will again focus on the Earth Summit. When Governments meet in New York for the UN General Assembly's special session to review progress since Rio, the question will be: What changes have the major players - including Governments, international policy makers, businesses, trade unions, farmers and women's groups - been able to bring about in the five years since Rio? A great deal has happened, but, in the view of some, not nearly enough to achieve the Summit's goals. There is growing awareness of the many "negative incentives" which continue to encourage people to become wasteful consumers. The Commission intends to elaborate for the 1997 special session of the GA concrete proposals for mechanisms and policy instruments to facilitate achieving the aims of Rio.

— Rio Declaration on Environment and Development

— The United Nations Conference on Environment and Development,

— Having met at Rio de Janeiro from 3 to 14 June 1992,

— Reaffirming the Declaration of the United Nations Conference on the Human Environment, adopted at Stockholm on 16 June 1972, and seeking to build upon it,

— With the goal of establishing a new and equitable global partnership through the creation of new levels of cooperation among States, key sectors of societies and people,

— Working towards international agreements which respect the interests of all and protect the integrity of the global environmental and developmental system,

— Recognizing the integral and interdependent nature of the Earth, our home,

Proclaims that:

Principle 1

Human beings are at the centre of concerns for sustainable

development. They are entitled to a healthy and productive life in harmony with nature.

Principle 2

States have, in accordance with the Charter of the United Nations and the principles of international law, the sovereign right to exploit their own resources pursuant to their own environmental and developmental policies, and the responsibility to ensure that activities within their jurisdiction or control do not cause damage to the environment of other States or of areas beyond the limits of national jurisdiction

Principle 3

The right to development must be fulfilled so as to equitably meet developmental and environmental needs of present and future generations.

Principle 4

In order to achieve sustainable development, environmental protection shall constitute an integral part of the development process and cannot be considered in isolation from it.

Principle 5

All States and all people shall cooperate in the essential task of eradicating poverty as an indispensable requirement for sustainable development, in order to decrease the disparities in standards of living and better meet the needs of the majority of the people of the world.

Principle 6

The special situation and needs of developing countries, particularly the least developed and those most environmentally vulnerable, shall be given special priority. International actions in the field of environment and development should also address the interests and needs of all countries.

Principle 7

States shall cooperate in a spirit of global partnership to conserve, protect and restore the health and integrity of the Earth's ecosystem. In view of the different contributions to global environmental degradation, States have common but differentiated responsibilities. The developed countries acknowledge the responsibility that they bear in the international pursuit to sustainable development in view of the pressures their societies place on the global environment and of the technologies and financial resources they command.

Principle 8

To achieve sustainable development and a higher quality of life for all people, States should reduce and eliminate unsustainable patterns of production and consumption and promote appropriate demographic policies.

Principle 9

States should cooperate to strengthen endogenous capacity-building for sustainable development by improving scientific understanding through exchanges of scientific and technological knowledge, and by enhancing the development, adaptation, diffusion and transfer of technologies, including new and innovative technologies.

Principle 10

Environmental issues are best handled with participation

of all concerned citizens, at the relevant level. At the national level, each individual shall have appropriate access to information concerning the environment that is held by public authorities, including information on hazardous materials and activities in their communities, and the opportunity to participate in decision-making processes. States shall facilitate and encourage public awareness and participation by making information widely available. Effective access to judicial and administrative proceedings, including redress and remedy, shall be provided.

Principle 11

States shall enact effective environmental legislation. Environmental standards, management objectives and priorities should reflect the environmental and development context to which they apply. Standards applied by some countries may be inappropriate and of unwarranted economic and social cost to other countries, in particular developing countries.

Principle 12

States should cooperate to promote a supportive and open international economic system that would lead to economic growth and sustainable development in all countries, to better address the problems of environmental degradation. Trade policy measures for environmental purposes should not constitute a means of arbitrary or unjustifiable discrimination or a disguised restriction on international trade.

Unilateral actions to deal with environmental challenges outside the jurisdiction of the importing country should be avoided. Environmental measures addressing transboundary or global environmental problems should, as far as possible, be based on an international consensus.

Principle 13

States shall develop national law regarding liability and compensation for the victims of pollution and other environmental damage. States shall also cooperate in an expeditious and more determined manner to develop further international law regarding liability and compensation for adverse effects of environmental damage caused by activities within their jurisdiction or control to areas beyond their jurisdiction.

Principle 14

States should effectively cooperate to discourage or prevent the relocation and transfer to other States of any activities and substances that cause severe environmental degradation or are found to be harmful to human health.

Principle 15

In order to protect the environment, the precautionary approach shall be widely applied by States according to their capabilities. Where there are threats of serious or irreversible damage, lack of full scientific certainty shall not be used as a reason for postponing cost-effective measures to prevent environmental degradation.

Principle 16

National authorities should endeavour to promote the internalization of environmental costs and the use of economic instruments, taking into account the approach that the polluter should, in principle, bear the cost of pollution, with due regard to the public interest and without distorting international trade and investment.

Principle 17

Environmental impact assessment, as a national

instrument, shall be undertaken for proposed activities that are likely to have a significant adverse impact on the environment and are subject to a decision of a competent national authority.

Principle 18

States shall immediately notify other States of any natural disasters or other emergencies that are likely to produce sudden harmful effects on the environment of those States. Every effort shall be made by the international community to help States so afflicted.

Principle 19

States shall provide prior and timely notification and relevant information to potentially affected States on activities that may have a significant adverse transboundary environmental effect and shall consult with those States at an early stage and in good faith.

Principle 20

Women have a vital role in environmental management and development. Their full participation is therefore essential to achieve sustainable development.

Principle 21

The creativity, ideals and courage of the youth of the world should be mobilized to forge a global partnership in order to achieve sustainable development and ensure a better future for all.

Principle 22

Indigenous people and their communities and other local communities have a vital role in environmental management and development because of their knowledge and traditional practices. States should recognize and duly support their identity, culture and interests and enable their effective participation in the achievement of sustainable development.

Principle 23

The environment and natural resources of people under oppression, domination and occupation shall be protected.

Principle 24

Warfare is inherently destructive of sustainable development. States shall therefore respect international law providing protection for the environment in times of armed conflict and cooperate in its further development, as necessary.

Principle 25

Peace, development and environmental protection are interdependent and indivisible.

Principle 26

States shall resolve all their environmental disputes peacefully and by appropriate means in accordance with the Charter of the United Nations.

Principle 27

States and people shall cooperate in good faith and in a spirit of partnership in the fulfilment of the principles embodied in this Declaration and in the further development of international law in the field of sustainable development.

11

Declaration of the United Nations Conference on the Human Environment- 1972

The United Nations Conference on the Human Environment, having met at Stockholm from 5 to 16 June 1972, having considered the need for a common outlook and for common principles to inspire and guide the peoples of the world in the preservation and enhancement of the human environment, proclaims that:

Man is both creator and moulder of his environment, which gives him physical sustenance and affords him the opportunity for intellectual, moral, social and spiritual growth. In the long and tortuous evolution of the human race on this planet a stage has been reached when, through the rapid acceleration of science and technology, man has acquired the power to transform his environment in countless ways and on an unprecedented scale.

Both aspects of man's environment, the natural and the manmade, are essential to his well-being and to the enjoyment of basic human rights - even the right to life itself. The protection and improvement of the human environment is a major issue which affects the well-being of peoples and economic development throughout the world; it is the urgent desire of the peoples of the whole world and the duty of all Governments.

Man has constantly to sum up experience and go on discovering, inventing, creating and advancing. In our time, man's capability to transform his surroundings, if used wisely, can bring to all peoples the benefits of development and the opportunity to enhance the quality of life. Wrongly or heedlessly applied, the same power can do incalculable harm to human beings and the human environment.

We see around us growing evidence of man-made harm in many regions of the earth: dangerous levels of pollution in water, air, earth and living beings; major and undesirable disturbances to the ecological balance of the biosphere; destruction and depletion of irreplaceable resources; and gross deficiencies, harmful to the physical, mental and social health of man, in the man-made environment, particularly in the living and working environment.

In the developing countries most of the environmental problems are caused by under-development. Millions continue to live far below the minimum levels required for a decent human existence, deprived of adequate food and clothing, shelter and education, health and sanitation. Therefore, the developing countries must direct their efforts to development, bearing in mind their priorities and the need to safeguard and improve the environment.

For the same purpose, the industrialized countries should make efforts to reduce the gap between themselves and the developing countries. In the industrialized countries, environmental problems are generally related to industrialization and technological development. The natural growth of population continuously presents problems for the preservation of the environment, and adequate policies and measures should be adopted, as appropriate, to face these problems.

Of all things in the world, people are the most precious. It is the people that propel social progress, create social wealth, develop science and technology and, through their hard work, continuously transform the human environment. Along with social progress and the advance of production, science and technology, the capability of man to improve the environment increases with each passing day.

A point has been reached in history when we must shape our actions throughout the world with a more prudent care for their environmental consequences. Through ignorance or indifference we can do massive and irreversible harm to the earthly environment on which our life and well-being depend. Conversely, through fuller knowledge and wiser action, we can achieve for ourselves and our posterity a better life in an environment more in keeping with human needs and hopes.

There are broad vistas for the enhancement of environmental quality and the creation of a good life. What is needed is an enthusiastic but calm state of mind and intense but orderly work. For the purpose of attaining freedom in the world of nature, man must use knowledge to build, in collaboration with nature, a better environment. To defend and improve the human environment for present and future generations has become an imperative goal for mankind - a goal to be pursued together with, and in harmony with, the established and fundamental goals of peace and of world-wide economic and social development.

To achieve this environmental goal will demand the acceptance of responsibility by citizens and communities and by enterprises and institutions at every level, all sharing equitably in common efforts. Individuals in all walks of life as well as organizations in many fields, by their values and the sum of their actions, will shape the world environment of the future. Local and national governments will bear the greatest burden for large-scale environmental policy and action within their jurisdictions. International co-operation is also needed in order to raise resources to support the developing countries in carrying out their responsibilities in this field.

A growing class of environmental problems, because they are regional or global in extent or because they affect the common international realm, will require extensive co-operation among nations and action by international organizations in the common interest. The Conference calls upon Governments and peoples to exert common efforts for the preservation and improvement of the human environment for the benefit of all the people and for their posterity.

PRINCIPLES

The United Nations Conference on the Human Environment states the common conviction that:

Principle 1

Man has the fundamental right to freedom, equality and adequate conditions of life, in an environment of a quality that permits a life of dignity and well-being, and he bears a solemn responsibility to protect and improve the environment for present and future generations. In this respect, policies promoting or perpetuating apartheid, racial segregation, discrimination, colonial and other forms of oppression and foreign domination stand condemned and must be eliminated.

Principle 2

The natural resources of the earth, including the air, water, land, flora and fauna and especially representative samples of natural ecosystems, must be safeguarded for the benefit of present and future generations through careful planning or management, as appropriate.

Principle 3

The capacity of the earth to produce vital renewable resources must be maintained and, wherever practicable, restored.

Principle 4

Man has special responsibility to safeguard and wisely manage the heritage of wildlife and its habitat which are now gravely imperilled by a combination of adverse factors. Nature conservation, including wildlife, must therefore receive importance in planning for economic development.

Principle 5

The non-renewable resources of the earth must be employed in such a way as to guard against the danger of their future exhaustion and to ensure that benefits from such employment are shared by all mankind.

Principle 6

The discharge of toxic substances or of other substances and the release of heat, in such quantities or concentrations as to exceed the capacity of the environment to render them harmless, must be halted in order to ensure that serious or irreversible damage is not inflicted upon ecosystems. The just struggle of the peoples of all countries against pollution should be supported.

Principle 7

States shall take all possible steps to prevent pollution of the seas by substances that are liable to create hazards to human health, to harm living resources and marine life; to damage amenities or to interfere with other legitimate uses of the sea.

Principle 8

Economic and social development is essential for ensuring a favourable living and working environment for man and for creating conditions on earth that are necessary for the improvement of the quality of life.

Principle 9

Environmental deficiencies generated by the conditions of under-development and natural disasters pose grave problems and can best be remedied by accelerated development through the transfer of substantial quantities of financial and technological assistance as a supplement to the domestic effort of the developing countries and such timely assistance as may be required.

Principle 10

For the developing countries, stability of prices and adequate earnings for primary commodities and raw materials are essential to environmental management since economic factors as well as ecological processes must be taken into account.

Principle 11

The environmental policies of all States should enhance and not adversely affect the present or future development potential of developing countries, nor should they hamper the attainment of better living conditions for all, and appropriate steps should be taken by States and international organizations with a view to reaching agreement on meeting the possible national and international economic consequences resulting from the application of environmental measures.

Principle 12

Resources should be made available to preserve and improve the environment, taking into account the circumstances and particular requirements of developing countries and costs which may emanate from their incorporating environmental safeguards into their development planning and the need for making available to them, upon their request, additional international technical and financial assistance for this purpose.

Principle 13

In order to achieve a more rational management of resources and thus to improve the environment, States should adopt an integrated and coordinated approach to their development planning so as to ensure that development is compatible with the need to protect and improve the human environment for the benefit of their population.

Principle 14

Rational planning constitutes an essential tool for reconciling any conflict between the needs of development and the need to protect and improve the environment.

Principle 15

Planning must be applied to human settlements and urbanization with a view to avoiding adverse effects on the environment and obtaining maximum social, economic and environmental benefits for all. In this respect, projects which are designed for colonialist and racist domination must be abandoned.

Principle 16

Demographic policies which are without prejudice to basic human rights and which are deemed appropriate by Governments concerned should be applied in those regions where the rate of population growth or excessive population concentrations are likely to have adverse effects on the environment or development, or where low population densitiy may prevent improvement of the human environment and impede development.

Principle 17

Appropriate national institutions must be entrusted with the task of planning, managing or controlling the environmental resources of States with the view of enhancing environmental quality.

Principle 18

Science and technology, as part of their contribution to economic and social development, must be applied to the identification, avoidance and control of environmental risks and the solution of environmental problems and for the common good of mankind.

Principle 19

Education in environmental matters, for the younger generation as well as adults, giving due consideration to the underprivileged, is essential in order to broaden the basis for an enlightened opinion and responsible conduct by individuals, enterprises and communities in protecting and improving environment in its full human dimensions.

It is also essential that mass media of communication avoid contributing to the deterioration of the environment, but, on the contrary, disseminate information of an educational nature, on the need to protect and improve the environment in order to enable man to develop in every respect.

Principle 20

Scientific research and development in the context of environmental problems, both national and multinational, must be promoted in all countries, especially the developing countries. In this connexion, the free flow of up-to-date scientific information and transfer of experience must be supported and assisted, to facilitate the solution of environmental problems; environmental technologies should be made available to developing countries on terms which would encourage their wide dissemination without constituting an economic burden on the developing countries.

Principle 21

States have, in accordance with the Charter of the United Nations and the principles of international law, the sovereign right to exploit their own resources pursuant to their own environmental policies, and the responsibility to ensure that activities within their jurisdiction or control do not cause damage to the environment of other States or of areas beyond the limits of national jurisdiction.

Principle 22

States shall co-operate to develop further the international law regarding liability and compensation for the victims of pollution and other environmental damage caused by activities within the jurisdiction or control of such States to areas beyond their jurisdiction.

Principle 23

Without prejudice to such criteria as may be agreed upon by the international community, or to standards which will have to be determined nationally, it will be essential in all cases to consider the systems of values prevailing in each country, and the extent of the applicability of standards which are valid for the most advanced countries but which may be inappropriate and of unwarranted social cost for the developing countries.

Principle 24

International matters concerning the protection and improvement of the environment should be handled in a co-operative spirit by all countries, big or small, on an equal footing. Co-operation through multilateral or bilateral arrangements or other appropriate means is essential to effectively control, prevent, reduce and eliminate adverse environmental effects resulting from activities conducted in all spheres, in such a way that due account is taken of the sovereignty and interests of all States.

Principle 25

States shall ensure that international organizations play a co-ordinated, efficient and dynamic role for the protection and improvement of the environment.

Principle 26

Man and his environment must be spared the effects of nuclear weapons and all other means of mass destruction. States must strive to reach prompt agreement, in the relevant international organs, on the elimination and complete destruction of such weapons.

Bibliography

Callenbach, E., Capra, F., Goldman, L., Lutz, R., & Marburg, S. *EcoManagement: The Elmwood guide to ecological auditing and sustainable business*. San Francisco, CA: Berrett-Koehler, 1993.

Carson, P. & Moulden, J. *Green is gold: Business talking to business about the environmental revolution*. Toronto: HarperCollins Publ, 1991.

Elkington, J. *Towards the sustainable corporation: Win-win-win business strategies for sustainable development*. California Management Review, 36(2), 90-100, 1994.

Epstein, M. *Measuring corporate environmental performance*. Montvale, NJ: Institute of Management Accountants, 1995.

Fischer, K., & Schot, J. (Eds.) *Environmental strategies for industry: International perspectives on research needs and policy implications*. Washington,DC: Island Press, 1993.

Feyerherm, A.E. "Changing and converging mind-sets of participants during collaborative, environmental rule making: Two negotiated regulation case studies."

Halme, M., Environmental issues in product development processes: Paradigm shift in a Finnish packaging company. Business Ethics Quarterly, 5(4), 713-733, 1995.

Kohler, L.R. *Eco-auditing and environmental performance*. European Participation Monitor, 11, 37-40, 1995.

Lawrence, A.T., & Morell, D. "Leading-edge environmental management: Motivation, opportunity, resources, and processes."

Piasecki, G. Corporate environmental strategy: The avalanche of change since Bhopal. New York: John Wiley & Sons, 1995.

Piasecki, B., & Asmus, P. *In search of environmental excellence: Moving beyond blame.* Toronto: Simon & Schuster Inc., 1990.

Schlapkohl, *Environmental protection at BASF AG.* European Participation Monitor, 11, 27-30, 1995.

Starik, M. "Research on organizations and the natural environment: Some paths we have travelled, the "field" ahead."

Turcotte, M-F. "Conflict and collaboration: The interfaces between environmental organizations and business firms."

White, M.A. "The performance of environmental mutual funds in the United States and Germany: Is there economic hope for "green" investors?".

Winn, M.I. "Corporate leadership and policies for the natural environment."

Index